The Control and Stimulation of Follicular Growth

.

Advances
in Reproductive
Endocrinology

VOLUME 5

The Control and Stimulation of Follicular Growth

Edited by RW Shaw

The Parthenon Publishing Group

International Publishers in Medicine, Science & Technology

Casterton Hall, Carnforth,
Lancs, LA6 2LA, UK

One Blue Hill Plaza, Pearl River,
New York 10965, USA

Published in the UK by
The Parthenon Publishing Group Limited
Casterton Hall, Carnforth,
Lancs, LA6 2LA, England

Published in the USA by
The Parthenon Publishing Group Inc.
One Blue Hill Plaza
PO Box 1564, Pearl River,
New York 10965, USA

British Library Cataloguing in Publication Data
Control and Stimulation of Follicular
Growth. – (Advances in Reproductive
Endocrinology Series; Vol. 5)
  I. Shaw, Robert W. II. Series
  618.1

  ISBN   1-85070-470-8

Library of Congress Cataloging-in-Publication Data
The Control and stimulation of follicular growth / edited by R.W. Shaw
      p.   cm. — (Advances in reproductive endocrinology series ; v. 5)
  Includes bibliographical references and index.
  ISBN   1-85070-470-8
  1. Ovaries — Growth — Regulation.   2. Luteinizing hormone releasing
hormone — Physiological effect.   3. Follicle-stimulating hormone —
Physiological effect.   I. Shaw, Robert W. (Robert Wayne)
II. Series: Advances in reproductive endocrinology ; v. 5.
QP261.C68   1993
612.6'2—dc20   93-1595
                                                  CIP

Composition by Ryburn Typesetting, Keele University, England
Printed and bound in Great Britain by
Butler and Tanner Ltd, Frome and London

Contents

List of principal contributors   vii

Foreword   ix

1. Gonadotrophins and follicular growth   1
    B. Mannaerts and R. de Leeuw

2. Morphology of the human follicle during the
    preovulatory phase: normal and abnormal cases   13
    O. Bomsel-Helmreich

3. Local control of ovarian function   27
    S.G. Hillier

4. Follicular maturation in childhood and puberty   35
    I.A. Hughes

5. Ovulatory dysfunction in endocrine disorders   49
    P.G. Wardle and R. Fox

6. Purified gonadotrophin preparations for induction
    of ovulation   67
    *S. Franks, D.W. Polson, M. Sagle, D. Hamilton-Fairley,
    D.S. Kiddy and H.D. Mason*

7. Clinical aspects of recombinant human follicle
    stimulating hormone   75
    B. Mannaerts and H. Coelingh Bennink

v

8.   Management of ovulatory disorders with pulsatile
     gonadotrophin releasing hormone   87
     M. Filicori, G. Cognigni, L. Michelacci, P. Dellai,
     M. Sambataro and F. Carbone

9.   Growth hormone and ovarian stimulation   97
     H.S. Jacobs

10. Ovarian surgery   111
     A.Abdel Gadir

11. Premature luteinization   125
     R. Fleming, M.E. Jamieson and J.R.T. Coutts

12. Corpus luteal insufficiency   137
     M.G. Hammond

13. Multiple follicular maturation for assisted reproduction   151
     S.M. Walker

14. Risks associated with ovulation induction   163
     N. Amso

     Index   171

List of principal contributors

N. Amso
Newcastle General Hospital
Westgate Road
Newcastle upon Tyne NE4 6BE
UK

O. Bomsel-Helmreich
University Laboratory
Department of Obstetrics and
  Gynaecology
Hôpital A Belcère
157 Rue Porte de Trivaux
92141 Clamart
France

M. Filicori
Center for Chronobiology of
  Reproduction
University of Bologna
Via Massarenti 13
40138 Bologna
Italy

R. Fleming
Department of Obstetrics and
  Gynaecology
University of Glasgow
Royal Infirmary
Queen Elizabeth Building
10 Alexandra Parade
Glasgow G31 2ER
UK

S. Franks
Department of Obstetrics and
  Gynaecology
St Mary's Hospital Medical
  School
Paddington
London W2 1PG
UK

A. Abdel Gadir
The Hallam Centre
112 Harley Street
London W1N 1AF
UK

M.G. Hammond
Department of Obstetrics and
  Gynecology
The University of North Carolina
  at Chapel Hill
Chapel Hill
North Carolina 27599-7570
USA

S.G. Hillier
Reproductive Endocrinology
  Laboratory
University of Edinburgh
37 Chalmers Street
Edinburgh EH3 9EW
UK

I.A. Hughes
Department of Paediatrics
University of Cambridge School
  of Clinical Medicine
Addenbrooke's Hospital
Hills Road
Cambridge CB2 2QQ
UK

H.S. Jacobs
Reproductive Endocrinology
University College and Middlesex
  School of Medicine
Cobbold Laboratories
Mortimer Street
London W1N 8AA
UK

B. Mannaerts
Scientific Development Group
Oraganon International BV
PO Box 20
5340 BH Oss
The Netherlands

R.W. Shaw
Department of Obstetrics and
  Gynaecology
University of Wales College of
  Medicine
Heath Park
Cardiff CF4 4XN
UK

S.M. Walker
Assisted Reproduction Unit
Department of Obstetrics and
  Gynaecology
University of Wales College of
  Medicine
Heath Park
Cardiff CF4 4XN
UK

P.G. Wardle
University of Bristol
Department of Obstetrics and
  Gynaecology
St Michael's Hospital
Bristol BS2 8EG
UK

Foreword

Although for many generations we have been aware that the gonadotrophins luteinizing hormone (LH) and follicle stimulating hormone (FSH) play a vital role in the control of follicular growth and oocyte maturation, the precise mechanisms by which they work are unclear. The importance of local paracrine control within the ovary is becoming realized but factors that govern initiation of maturation of the primary follicles remain unclear. Newer therapeutic options have become available with the introduction to clinical practice of biosynthetic gonadotrophins, purified urinary gonadotrophin preparations, gonado-trophin hormone analogues and growth hormone augmentation for follicular maturation. All of these aspects were discussed in detail together with the complications and problems that can arise from inducing follicular growth when exogenous gonadotrophin preparations are administered.

We were fortunate to be able to bring together a group of distinguished basic scientists and clinicians involved in research and the clinical manage-ment of follicular growth at a workshop addressing these issues. This was the 5th in a series of International Workshops in Reproductive Endocrinology and was held at the Swallow Hotel, Bristol, in September 1992. The meeting was kindly sponsored by Zeneca Pharmaceuticals (formerly part of the ICI Group). This monograph contains the manuscripts from the major contributors at that meeting and clarifies the advances that have been made but also amplifies the areas where continued research is essential for increasing our basic understanding which has to be the basis for developing newer treatment protocols and approaches for the future.

Professor Robert W. Shaw
Department of Obstetrics and Gynaecology
University of Wales College of Medicine
Cardiff

1

Gonadotrophins and follicular growth

B. Mannaerts and R. de Leeuw

INTRODUCTION

The relative contribution of follicle stimulating hormone (FSH) and luteinizing hormone (LH) to folliculogenesis has been under investigation for many years. The very first report on animal experiments which showed the synergism of FSH and LH in producing oestrogen secretion was published in 1941[1]. Thereafter, conflicting data on the requirements of LH were reported, probably because FSH preparations with varying degrees of residual LH were used in tests. More recently, through the expression of FSH in mammalian host cells[2,3], pure human FSH which is devoid of LH activity has become available. Animal experiments using this recombinant human FSH (recFSH) have provided further insight to the specific actions of FSH during follicular growth and steroidogenesis. The aim of this chapter is to review the biochemical and biological properties of recFSH in comparison to FSH of natural sources and to outline some main findings from studies of hypophysectomized animals in which the efficacy of recFSH alone, and in the presence of complementary LH activity, was examined.

RECOMBINANT FSH CHARACTERISTICS

The recFSH (Org 32489, Organon International) under investigation was produced by a Chinese Hamster Ovary (CHO) cell line transfected with plasmids containing the two subunit genes encoding human FSH[3]. The stable integration of the human DNA into the chromosomal DNA of the host cell was confirmed by various techniques including *in situ* hybridization. Like the human pituitary, the CHO cell line produces intact, glycosylated human FSH that is secreted constantly into the culture medium, the latter being used as the source for its further isolation. The final, purified, freeze-dried, recFSH preparation has a biochemical purity of 99.9% and, accordingly, a high specific *in vivo* bioactivity (approximately 10 000 IU/mg protein). Previous comparative studies[4,5] on the biological properties of this pure recFSH demonstrated that its receptor binding activity and *in vitro* and *in vivo* efficacy, are comparable to those of FSH isolated from natural sources. However, in contrast to natural FSH preparations, recFSH is guaranteed free from other hormones, such as LH. Indeed, it has been demonstrated that recFSH lacks intrinsic LH activity, in that interaction of recFSH with the LH receptor is insufficient to induce adequate testosterone production, as demonstrated in an *in vitro* mouse Leydig cell bioassay[5]. In this model recFSH exhibited less than 0.025 mIU LH activity per IU recFSH. Assuming a comparable interaction of recFSH with the human LH receptor, it is unlikely that recFSH has any LH activity when applied in animal models, or for therapeutic purposes.

FSH HETEROGENEITY

Like natural FSH[6], recFSH exists in different molecular forms which differ in charge, relative abundance, receptor affinity, bioactivity and plasma half-life. The heterogeneity of FSH is due to differences in the amount and/or composition of the carbohydrate residues, in particular of sialic acid. Relatively acidic FSH isohormones, which are thought to be more heavily sialylated, exhibit a lower degree of receptor binding and *in vitro* bioactivity than more basic isohormones. On the other hand, sialylation prevents FSH degradation in the liver and excretion by the kidney[7]. Therefore, acidic isoforms remain in the circulation longer, resulting in a relative high *in vivo* bioactivity. For natural FSH preparations, the number and relative amount of each isohormone species depends on the source (pituitary, serum or urine), the age and endocrine status of the donor[8,9] and the isolation procedure applied for its purification. The charge heterogeneity of recFSH is mainly determined by the host cell-line chosen for its production. For FSH, a CHO cell line is an obvious choice since these cells are known to synthesize glycoproteins with oligosaccharides identical or closely related to those found in man[10,11].

Different isohormones of FSH may be separated according to their isoelectric point by means of isoelectrofocusing or chromatofocusing techniques. Fractionation studies, in which human FSH isoforms were separated by chromatofocusing, have been performed on crude[12] and purified pituitary FSH, circulating FSH[13], purified urinary FSH[14], and pure recFSH[5]. RecFSH (Org 32489) exhibited a distribution of FSH immunoactivity between pI 6.2 and 3 with 54% recovery between pI 4.1 and 4.7 (see Table 1 and Figure 1). In the latter pI region comparable amounts of FSH were recovered after chromatofocusing of urinary FSH (Metrodin®) or crude intrapituitary FSH. In comparison to urinary FSH, recFSH seems to contain less acidic isoforms with pI < 4.1 and more isoforms with pI > 4.7. Both recombinant and urinary FSH appear to be less heterogeneous than crude intrapituitary FSH derived from donors whose sex and age were not stated[12]. Interestingly, chromatofocusing of intrapituitary FSH revealed a relative high FSH activity in the salt fraction (pI < 3.8) and an additional small neutral fraction (pI 7.6–7.1). Whether all these additional isoforms are actually secreted by the pituitary gland, or represent intermediate products of FSH glycosylation, remains to be elucidated.

Chromatofocusing of serum has revealed that during the follicular and luteal phase of the normal menstrual cycle, the majority of circulating FSH forms have pI values < 4.8, whereas during midcycle, or after treatment with oestrogenic steroids, considerably more alkaline FSH forms (pI 6.0–4.8) are secreted by the pituitary[13]. In comparison to crude pituitary FSH, the highly purified pituitary International Standard (IS) 83/575 contains only acidic fractions (pI < 4.7). Although the recovery of FSH (58%) after chromatofocusing of IS 83/575 was relatively low, the data suggest that this preparation has lost its more basic isoforms (pI > 4.7) during purification. This isohormone distribution, however, seems in good agreement with the discrepancy between the declared *in vivo* bioactivity (80 IU/ampoule) and the actual, much lower, *in vitro* bioactivity (30 IU/ampoule) and immunoreactivity (16 IU/ampoule) of this preparation[15]. Moreover, it also explains the relatively high *in vivo* bioactivity : immunoreactivity ratio of this pituitary preparation in comparison to other FSH preparations[5].

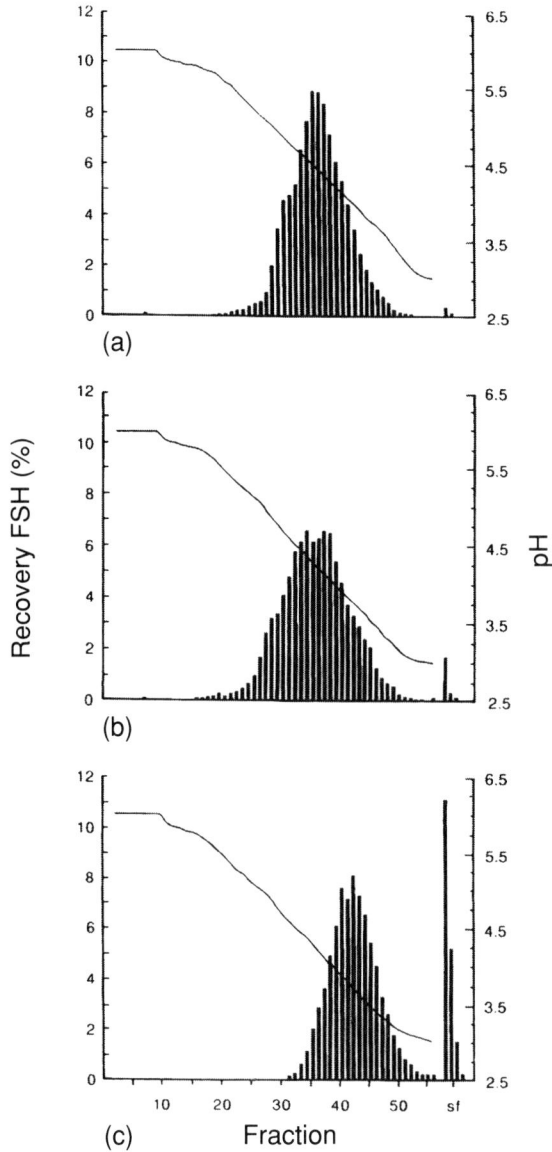

Figure 1 Immunoactivities of chromatofocusing fractions of (a) recombinant follicle stimulating hormone (FSH; Org 32489), (b) urinary FSH (Metrodin®) and (c) highly purified pituitary FSH (IS 83/575). Recoveries of FSH immunoactivities were 102, 78 and 58%, respectively; sf = high salt fraction

Table 1 Recovery (%) of eluted follicle stimulating hormone (FSH) immunoreactivity in different isohormone fractions originating from crude pituitary FSH (data from Ulloa Aguire and colleagues[12], with permission), recombinant FSH (Org 32489, batch 77), urinary FSH (Metrodin®, batch 07325089) and highly purified pituitary FSH (IS 83/575)

| | FSH | | | | Crude pituitary |
pH	Recombinant	Urinary	Pituitary	pH	FSH
—	—	—	—	7.6–7.1	1.5
6.2–5.3	3.5	1.7	0	5.9–5.3	8.9
5.3–4.7	29.9	19.9	0	5.0–4.7	14.4
4.7–4.1	54.1	51.5	11.3	4.5–4.1	54.8
4.1–3.6	10.9	22.8	41.7	3.9–3.8	3.7
3.6–3.0	1.6	3.1	29.0	—	—
< 3.0★	0.2	1.0	18.3	< 3.8★	16.8

★ Salt fraction

CONCEPTS OF FOLLICULAR DEVELOPMENT

Folliculogenesis is thought to comprise two phases[16,17]. First, primordial follicles are recruited and develop to small antral follicles. This initial growth is gonadotrophin-independent and continues after hypophysectomy. Whenever a follicle reaches the antral stage and the level of FSH exceeds a certain threshold value, the follicle continues its development to the preovulatory stage. However, if the antral stage is reached and FSH levels are insufficient, follicles become atretic. The total number of selected follicles would depend on the time period that FSH remains above the assumed threshold value. The exact role of LH during folliculogenesis is less well understood; developing follicles acquire their LH receptors only in the antral stage, in response to stimulation by FSH and oestradiol. *In vitro* experiments using rat granulosa cells indicate that low concentrations of LH stimulate aromatase and oestrogen production, whereas high LH concentrations impair oestrogen synthesis[18]. Biosynthesis of oestradiol requires both FSH and LH activity, since LH induces the synthesis of androgens, in the theca cell, which are

5

subsequently converted to oestrogens by aromatase enzymes induced by FSH in the granulosa cell[16]. Further proof which supports this so-called 'two-cell two-gonadotrophin' concept, as well as further insight into the regulatory function of FSH in follicle growth and atresia, has been obtained by testing recFSH in hypophysectomized animals.

FOLLICULAR DEVELOPMENT INDUCED BY RECOMBINANT FSH ONLY

In order to examine whether LH activity is indeed required for oestrogen synthesis, immature hypophysectomized rats were treated for 4 days, twice daily, with either vehicle solution, recFSH (total dosages 2.5, 5, 10, 20 and 40 IU), or urinary FSH (Metrodin, dosages as recFSH) containing minor amounts of LH activity (FSH : LH ratio ≥ 60)[5]. The main efficacy parameters were ovarian weight, ovarian aromatase activity and serum oestradiol levels. After treatment with recFSH, the total number of antral follicles (diameter $> 275\ \mu m$) and the incidence of atresia in these follicles were also evaluated, as described by Meijs-Roelofs and colleagues[19] and Osman[20]. Ovarian weight, ovarian aromatase activity and the total number of antral follicles increased dose-dependently after treatment with recFSH. Increases of ovarian weight and aromatase were comparable for recFSH and urinary FSH (Figure 2). Regardless of ovarian responses, serum oestradiol levels remained low at baseline (5–10 pg/ml) after treatment with either of these preparations, indicating that the remaining LH activity of the applied urinary FSH preparation was insufficient to support FSH-induced oestrogen secretion adequately. These data confirm the two-cell two-gonadotrophin concept that both FSH and LH are necessary for oestrogen biosynthesis.

Histological examination of the ovaries revealed that increasing recFSH doses caused a gradual shift of small antral follicles (Class 1) to large, pre-ovulatory follicles (Classes 4 and 5), whereas corpora lutea were not observed. The maturity of induced pre-ovulatory follicles was proven by means of a single human chorionic gonadotrophin (hCG) injection (10 IU) following treatment with 20 or 40 IU recFSH; this resulted, respectively, in 7 (of 8) and 8 (of 8) ovulating animals. Interestingly, the incidence of atresia diminished with increasing doses of recFSH. This correlation was most apparent in the smallest antral follicles (Class 1,

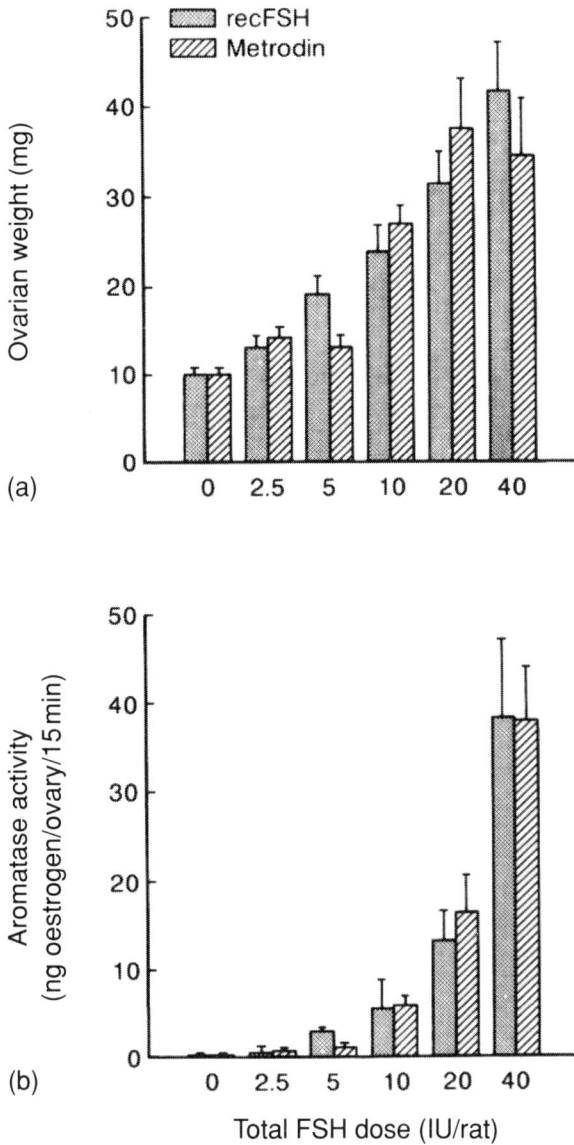

Figure 2 (a) Ovarian weight and (b) ovarian aromatase activity in immature hypophy-sectomized rats after treatment with increasing doses of recombinant follicle stimulating hormone (recFSH; Org 32489, batch 77) and urinary FSH (Metrodin®, batch 91B21)

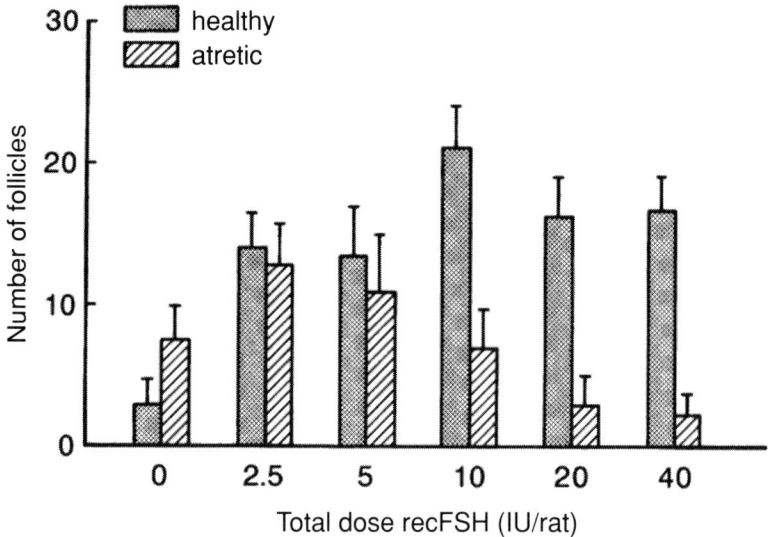

Figure 3 Number of small antral follicles (size Class 1, 275–350 μm diameter) classified as healthy and atretic after treatment with increasing doses of recombinant follicle stimulating hormone (recFSH; Org 32489, batch 65)

275–350 μm diameter) indicating that, at least in this animal model, FSH induces multiple follicular development by preventing small antral follicles from atresia rather than by activating additional follicle growth. The contribution of healthy and atretic follicles to the total number of antral follicles in the smallest size class is depicted in Figure 3.

FOLLICULAR DEVELOPMENT INDUCED BY recFSH AND hCG

In order to examine how much LH activity should be supplemented to FSH to increase plasma oestradiol levels in the animal model described above, additional experiments were performed using 40 IU recFSH which is known to induce maximal responses of aromatase activity[5]. RecFSH supplemented with 0.1, 1 or 10 IU hCG induced quickly rising serum oestradiol levels and the lowest dose (0.1 IU hCG) caused a threefold increase of circulating oestradiol from 10 to 34 pg/ml, indicating that the amount of LH activity required to support oestrogen biosynthesis is very low. In other experiments, animals were treated with a total of 8 IU recFSH supplemented with 0.2, 0.5, 2 or 5 IU hCG or with 5 IU hCG only. Addition of hCG increased ovarian weight in a dose-dependent fashion, but no further increases of the total number of follicles were noted. Nevertheless, addition of only 0.2 IU hCG caused a considerable shift of small follicles (Classes 1 and 2) to large antral follicles (Classes 4 and 5) and higher doses of hCG provided a comparable size distribution. Reduction of atresia was noted in antral follicles of all size-classes, especially after treatment with 0.5 IU hCG.

SUMMARY

In comparison with natural follicle stimulating hormone preparations, recombinant human FSH (recFSH, Org 32489) has a very high purity (99.9%) and lacks intrinsic luteinizing hormone activity. Like natural FSH, recFSH exhibits a considerable charge heterogeneity with a distribution of FSH isohormones between pI 6.2 and 3.0.

Recombinant FSH was administered to immature, female hypophysectomized rats to elucidate concepts of follicular growth, atresia and steroidogenesis. In this animal model, FSH induced normal follicular growth up to the preovulatory stage, whereas circulating oestradiol levels remained low at the baseline.

Ovarian weight, aromatase activity and the number of antral follicles increased with the FSH dose given, whereas the latter was negatively correlated with the incidence of atresia in the smallest size-class of antral follicles. Supplementation of recFSH with human chorionic gonadotrophin revealed that minute amounts of LH activity largely increase circulating oestradiol levels. Human chorionic gonadotrophin also augmented FSH-induced increases of ovarian weight and aromatase activity, but did not cause a further increase in the number of antral follicles. The incidence of atresia was decreased in all size classes.

It is concluded that FSH alone is able to induce follicle growth up to the preovulatory stage, but that the amount of LH activity influences the percentage of healthy follicles.

ACKNOWLEDGEMENTS

The authors would like to thank the Department of Endocrinology and Reproduction, Erasmus University, Rotterdam, The Netherlands, for performing the histotechnical work and especially Dr J. Uilenbroek for evaluating the histological data.

REFERENCES

1. Fevold, H.L. (1941). Synergism of follicle stimulating and luteinizing hormone in producing estrogen secretion. *Endocrinology*, **28**, 33–6
2. Keene, J.L., Matzuk, M.M., Otani, T., Fauser, B.C., Galway, A.B., Hsueh, A.J. and Boime, I. (1989). Expression of biologically active human follitropin in Chinese hamster ovary cells. *J. Biol. Chem.*, **246**, 4769–75
3. Van Wezenbeek, P., Draaier, J., Van Meel, F. and Olijve, W. (1990). Recombinant follicle stimulating hormone I. Construction, selection and characterization of a cell line. In Crommelin, D.J.A. and Schellekens, H., (eds.) *From Clone to Clinic, Developments in Biotherapy*, vol. 1, pp. 245–51. (London: Kluwer)
4. De Boer, W. and Mannaerts, B. (1990). Recombinant follicle stimulating hormone II. Biochemical and biological characteristics. In Crommelin, D.J.A. and Schellekens, H., (eds.) *From Clone to Clinic, Developments in Biotherapy*, vol. 1, pp. 253–9 (London: Kluwer)
5. Mannaerts, B., De Leeuw, R., Geelen, J., Van Ravenstein, A., Van Wezenbeek, P., Schuurs, A. and Kloosterboer, H. (1991). Comparative *in vitro* and *in vivo* studies on the biological properties of recombinant human follicle stimulating hormone. *Endocrinology*, **129**, 2623–30
6. Ulloa-Aguirre, A., Espinoza, R., Damian-Matsumura, P. and Chappel, S.C. (1988). Immunological and biological potencies of different molecular species of gonadotropins. *Hum. Reprod.*, **3**, 491–501
7. Wide, L. (1986). The regulation of metabolic clearance rate of human FSH in mice by variation of the molecular structure of the hormone. *Acta Endocrinol.*, **112**, 336–44
8. Wide, L. (1981). Male and female forms of human follicle-stimulating hormone in serum. *J. Clin. Endocrinol. Metab.*, **55**, 682–8
9. Wide, L. and Hobson, B.M. (1983). Qualitative differences in follicle-stimulating hormone activity in the pituitaries of young women compared to that of men and elderly women. *J. Clin. Endocrinol. Metab.*, **56**, 371–5
10. Sasaki, H., Bothner, B., Dell, A. and Fukuda, M. (1987). Carbohydrate

structure of erythropoietin expressed in Chinese hamster ovary cells by a human erythropoietic cDNA. *J. Biol. Chem.*, **262**, 12059–76

11. Hård, K., Mekking, A., Damm, J.B., Kamerling, J.P., Boer, W., Wijnands, R.A. and Vliegenthart, J.F. (1990). Isolation and structure determination of the intact sialylated N-linked carbohydrate chains of recombinant human follitropin (hFSH) expressed in Chinese hamster ovary cells. *Eur. J. Biochem.*, **193**, 263–71

12. Ulloa-Aguirre, A., Cravioto, A., Damián-Matsumura, P., Jiménez, M., Zambrano, E. and Díaz-Sánchez, V. (1992). Biological characterization of the naturally occurring analogues of intrapituitary human follicle stimulating hormone. *Hum. Reprod.*, **7**, 23–30

13. Padmanabhan, V., Lang, L.L., Sonstein, J., Kelch, R.P. and Beitins, I.Z. (1988). Modulation of serum follicle-stimulating hormone bioactivity and isoform distribution by estrogenic steroids in normal women and in gonadal dysgenesis. *J. Clin. Endocrinol. Metab.*, **67**, 465–73

14. Harlin, J., Khan, S.A. and Diczfalusy, E. (1986). Molecular composition of luteinizing hormone and follicle-stimulating hormone in commercial gonadotropin preparations. *Fertil. Steril.*, **46**, 1055–61

15. Storring, P.L. and Gaines Das, R.E. (1989). The International Standard for pituitary FSH: collaborative study of the standard and of four other purified human FSH preparations of differing molecular composition by bioassays, receptor assays and different immunoassay systems. *J. Endocrinol.*, **123**, 275–93

16. Baird, D.T. (1987). A model for follicular selection and ovulation: lessons from superovulation. *J. Steroid Biochem.*, **27**, 15–23

17. Baird, D.T. (1991). The ovarian cycle. In Hillier, G. (ed.), *Ovarian Endocrinology*, pp. 1–24 (Oxford: Blackwell Scientific Publications)

18. Overes, H.W.T.M., de Leeuw, R. and Kloosterboer, H.J. (1992). Regulation of aromatase activity in FSH-primed rat granulosa cells *in vitro* by follicle-stimulating hormone and various amounts of human chorionic gonadotrophin. *Hum. Reprod.*, **7**, 191–6

19. Meijs-Roelofs, H.M.A., Osman, P. and Kramer, P. (1982). Ovarian follicular development leading to first ovulation and accompanying gonadotrophin levels as studied in the unilaterally ovariectomized rat. *J. Endocrinol.*, **92**, 341–49

20. Osman, P. (1985). Rate and course of atresia during follicular development in the adult cyclic rat. *J. Reprod. Fertil.*, **73**, 261–70

2

Morphology of the human follicle during the preovulatory phase: normal and abnormal cases

O. Bomsel-Helmreich

INTRODUCTION

Our knowledge of the morphology of human mature oocytes at the time of retrieval for *in vitro* fertilization, i.e. 4–6 h before ovulation, is relatively detailed. Much less is known about the whole follicle over this period, and even less about the follicle before or during the late stages of follicular growth and during the time of preovulatory maturation, 0–38 h after human chorionic gonadotrophin (hCG) administration. This absence of information induced us to study the large and medium follicles obtained after several stimulation protocols (clomiphene citrate with human menopausal gonadotrophin (hMG), hMG or follicle stimulating hormone (FSH) alone, and gonadotrophin releasing hormone (GnRH) agonist combined with either hMG or FSH), followed possibly by a spontaneous luteinizing hormone (LH) surge or administration of hCG.

The elements of the follicle were observed by electron microscopy and the follicular fluid was assayed for steroids. Data obtained from 150 follicles and oocytes permitted a chronological description of the late follicular phase.

THE DOMINANT FOLLICLE BEFORE THE LH RISE OR ADMINISTRATION OF hCG, AT TIME 0

More than 10 years ago, Di Zerega and Hodgen[1] described recruitment and selection amongst the large follicles in the early follicular phase and the emergence of a dominant follicle, which induces atresia of the cohort follicles and the discontinuation of recruitment of any smaller follicles.

These are the events during the spontaneous cycle; in the case of ovarian stimulation, the stimulation process allows the rescue of the cohort follicles and, consequently, induces superovulation. Very few studies of large follicles during the follicular phase exist, however. Follow-up by ultrasound does not permit easy identification of the growth of individual follicles. The few existing studies show that the dominant follicle selected does not remain necessarily dominant until ovulation, but may well become atretic and be replaced by the next largest one (Bomsel-Helmreich[2] and references therein). Furthermore, recent studies of early embryonic loss allowed Boklage[3] to suggest that one natural conception in eight is a dizygotic twin, resulting from two preovulatory follicles which had grown and ovulated together. Only in the majority of spontaneous cycles does a single follicle become dominant and ovulate.

The dominant follicle near time 0, just before the LH rise or hCG administration, can be defined by LH receptor activity, a strong aromatase activity in the granulosa cells, 17α-hydroxylase activity in the theca interna, a high inhibin production and a high oestradiol : androgen ratio.

The morphology of the dominant follicle changes profoundly during the preovulatory phase, as shown in Figure 1. At 0 h, vascularization is already significant, which allows the follicular fluid to increase and a high mitotic activity in the granulosa (0.5–1.5%) is observed. During the final rise of oestrogen before the LH surge the granulosa retains a typical columnar organization, with a belt of a large nuclei near the basement membrane. Red blood cells are observed between the theca interna and the basement membrane, subsequent to the high vascularization of the theca interna. This follicular organization is observed in spontaneous cycles and also after most types of stimulation.

The treatment with GnRH agonists, however, induces a premature dissociation of the granulosa; Call Exner bodies, which are observed occasionally in spontaneous cycles, are very abundant, whatever their physiological role. At this period, the cells surrounding the oocyte, the

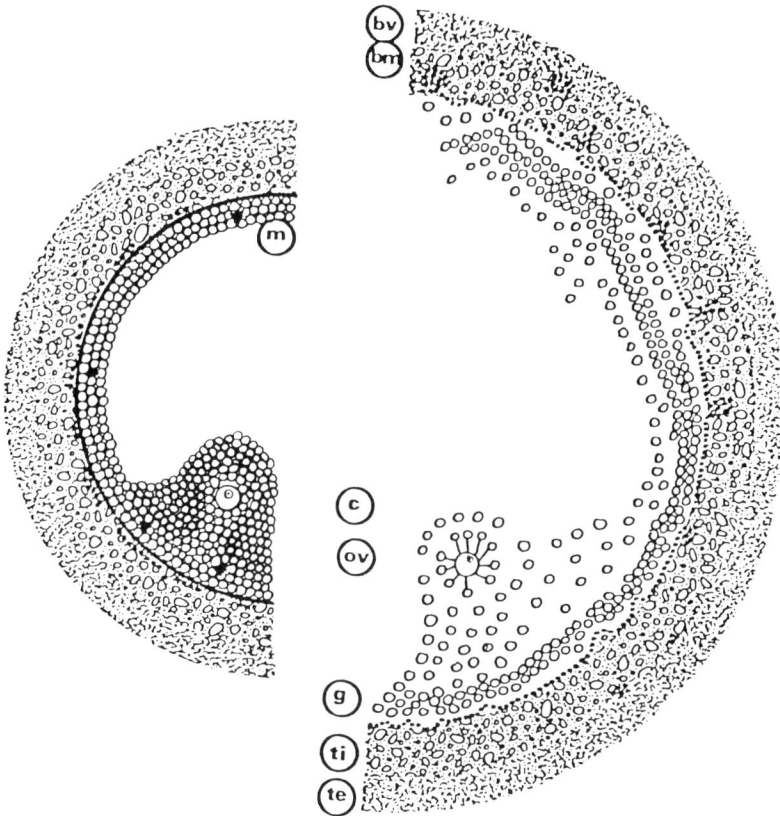

Figure 1 Progression of the preovulatory follicle just before the luteinizing hormone surge at 0 h (left) to the preovulatory follicle before ovulation, after 36 h (right). bv, blood vessel; bm, basal membrane; m, mitosis; c, cumulus-corona; ov, ovum; g, granulosa; ti, theca interna; te, theca externa

cumulus, and those closest to the oocyte (the corona cells) are round and regular, dividing at the same rate as the mural granulosa. However, some lacunae have already appeared in the proximity of the oocyte (Figure 2). Foot-processes of the corona cells connect them with the ooplasm by numerous gap-junctions after crossing the zona pellucida.

These junctions, seen in Figure 3, permit the transit of quite large metabolites, such as cyclic adenosine monophosphate, which plays an

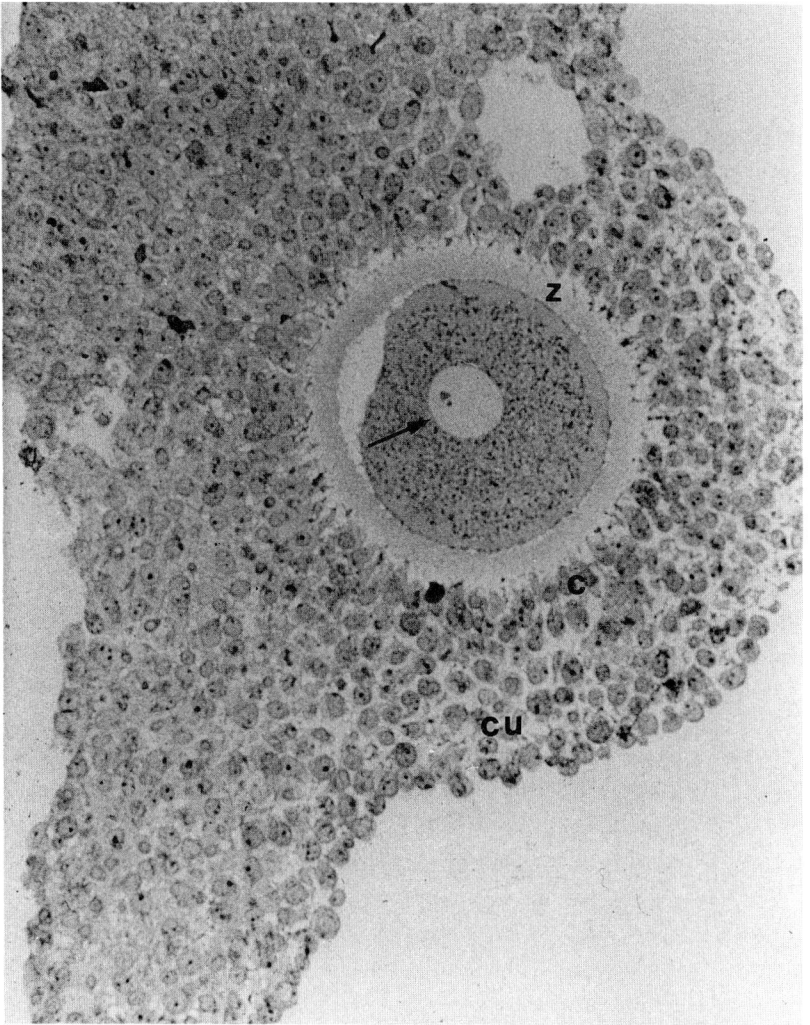

Figure 2 Human oocyte at time 0. The arrow shows the nucleus in germinal vesicle stage. cu, dense cumulus; c, corona; z, zona pellucida. (x 275)

essential role in the nuclear maturation process, inhibiting the resumption of meiosis.

The communication between oocyte and milieu is not a one-way system: the oocyte transmits signals also to the granulosa[4].

16

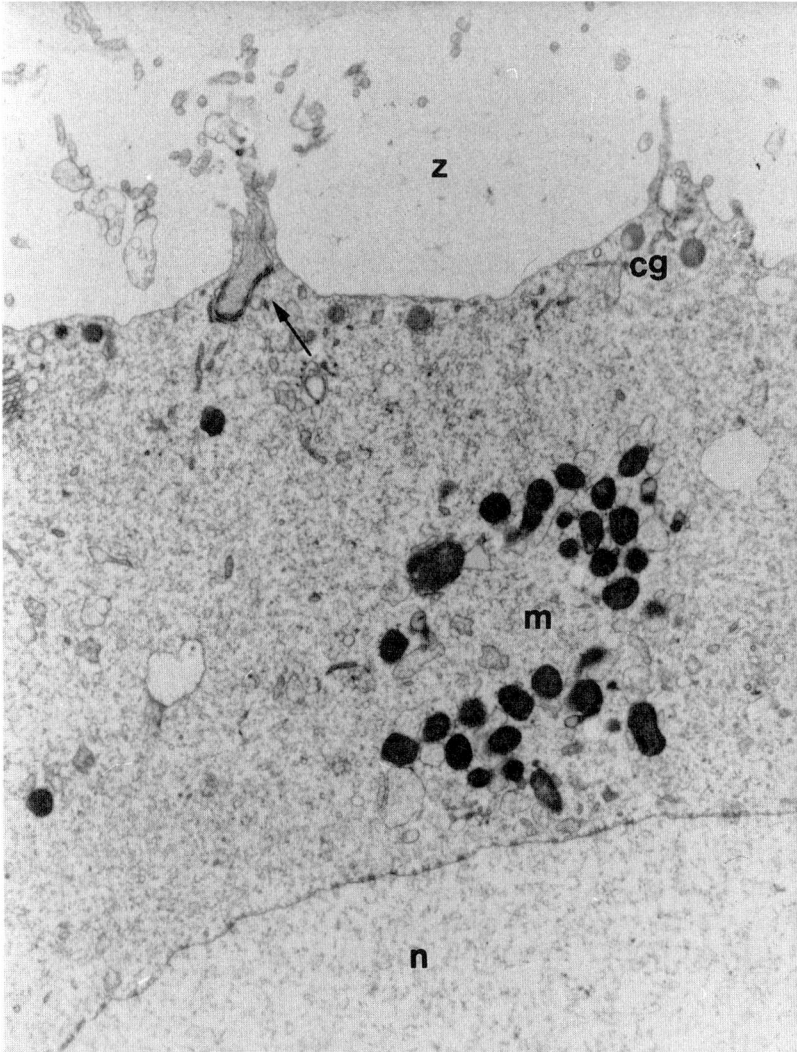

Figure 3 Human oocyte at time 0 (electron micrograph). The arrow shows a foot-process with junction. z, zona pellucida; cg, cortical granule; m, cluster of mitochondria in organelle-free cytoplasm; n, nucleus. (x 12 650)

The glycoprotein inert zona pellucida is 12–15 μm thick and formed by interlinked filaments which lie mostly parallel to the surface. The foot processes can penetrate more or less deeply into the ooplasm but mainly remain subcortical or superficial. Tight junctions are also observed[5]. The plasma membrane of the oocyte is covered with irregular microvilli. In its cortical region, a mixed population of cortical granules of two types – less dense (immature) and dense (mature ones) – form small clusters or an irregular layer intermixed with mitochondria.

During this period, a polarization of the organelles of the ooplasm is observed. The mitochondria, which are mostly round and the smooth endoplasmic reticulum, both of which are firmly associated with each other, are dispersed over the ooplasm, with the exception of a subcortical organelle-free layer underneath the membrane and the cortical granule layer, and another, perinuclear one. The round nucleus, also called the germinal vesicle, retreats from the central position observed in antral follicles to a peripheral position. The nuclear membrane is smooth. One large nucleolus is surrounded by an abundant mass of heterochromatin. This organization of the germinal vesicle is typical of the oocyte at 0 h. However, some nuclear maturation has already occurred since the small antral follicle phase, because the oocyte, at this earlier stage, does not show an intranuclear heterochromatin mass, but chromatin threads, which are visible from the primary oocyte stage.

THE FOLLICLE DURING THE INTERMEDIATE STAGE

This stage refers to the follicle during the rise of LH or after the administration of hCG, and before the breakdown of the germinal vesicle, 0–15 h after hCG.

Three main features must change in the preovulatory follicle to allow the achievement of oocyte maturation and ovulation: the dissociated granulosa and cumulus liberate the oocyte from the follicular wall, meiosis resumes in the oocyte and the follicle ruptures to allow the escape of the oocyte into the Fallopian tube. A chronology of the resumption of meiosis in the human oocyte has been established from other studies[6]. In the follicle, hyperaemia is induced by histamines and prostaglandin E_2. The progressive increase of the follicular diameter is caused by the increase of follicular fluid and not by an increase of the cell number as the mitotic index is lowered to 0.2–0.3%. The typical granulosa organization starts to disintegrate, cells and nuclei grow in size; progesterone secretion rises and cells start to pre-luteinize. In the follicular fluid, the high oestradiol : androgen ratio is progressively replaced by a high proges-terone : oestradiol ratio, which peaks at ovulation. The cumulus cells secrete hyaluronic acid which induces a dissociation of the cellular mass. Until 15 h post-hCG administration, the time of germinal vesicle breakdown, this is relatively discrete; corona cells elongate but a few mitoses are still seen. The corona cell foot-processes remain in contact with the ooplasm.

At 14 h post-hCG administration the nuclear membrane starts to undulate, then to invaginate. Heterochromatin disappears gradually and small chromosome-like elements condense; the nucleolus is the last to disappear. At 15 h, the germinal vesicle breaks down and disappears.

The stimulation by GnRH agonists, in contrast to classical stimulations with clomiphene citrate/hMG or FSH alone, induces specific changes before the administration of hCG or the spontaneous LH rise. Whereas most of the morphological characteristics remain the same as in otherwise stimulated oocytes, some elements indicate a tendency towards premature resumption of meiosis: cumulus cells dissociate and corona cells elongate, corresponding to 12–15 h post-hCG administration. Cortical granules migrate, mostly to a peripheral position, where they form a monolayer. The changes in the nucleus are even more striking. Often, the germinal vesicle has already migrated to an excentric position, as observed otherwise at 18 h after hCG administration, just before Metaphase I; the nuclear membrane is undulating, then invaginated, heterochromatin fades progressively and disappears, whereas chromosome-like condensed chromatin appears. These nuclear transformations resemble those observed 14 h, or even later[5], after hCG administration, with other stimulations.

FROM GERMINAL VESICLE BREAKDOWN TO OVULATION

The time interval discussed here is 15–40 h after hCG administration or until retrieval. The growth of the follicle increases notably in the hours preceding ovulation. The theca interna becomes oedematous by the blood flux induced by the increase of platelet activating factor and the basement membrane disappears. The granulosa shows almost no mitoses, fenestrated capillaries and red blood corpuscles are observed in the luteinizing cell mass. The increase in follicular fluid volume facilitates the dissociation of the granulosa, but this is incomplete, probably because of the local production of inhibitors of collagenase.

The increase of angiogenic factors in the follicular fluid and their effects, together with diverse enzymatic activities, suggests that the mechanism of ovulation should be considered as an inflammatory process[7]. The dissociation of the granulosa after the rise of LH in the spontaneous cycle shows a typical brush-like aspect, whereas after the administration of hCG, heavy preluteinization sets in with a smoother aspect. The dissociation of the cumulus is much more complete, because of its capacity to produce hyaluronic acid, a cell mass dissociation factor; the corona cells which are still linked to the ooplasm remain in place. A very high degree of this dissociation is already achieved at 20 h post-hCG administration, so that this dispersed cumulus becomes visible by ultrasound. The corona is remarkably elongated and gap junctions between corona cells are very rare. The zona pellucida becomes progressively thinner: this thinning proceeds with meiosis; it continues after fertilization until hatching of the blastocyst[8].

The foot processes of the corona draw away from the ooplasm after germinal vesicle breakdown; gap junctions progressively disappear, so that the connections between the follicular milieu and the oocyte are much reduced and the oocyte becomes isolated, which is necessary for its nuclear maturation. However, the presence of numerous coated vesicles in the cortical ooplasm suggests a possible continuation of traffic between the oocyte and its surroundings.

The microvilli of the plasma membrane become short and straight. Cortical granules form an irregular monolayer close to the membrane. Some of them extrude their contents all along the maturation process, without any atretic significance. They play an essential role in regulating fertilization. After penetration of the first sperm into the ooplasm, all granules immediately empty themselves into the perivitelline space; both the surface of the ooplasm and the zona pellucida hardens, consequently guaranteeing monospermy.

After germinal vesicle breakdown, the cytoplasmic polarization of the oocyte, with its two organelle-free regions, disappears; organelles are distributed all over the ooplasm. However, they do not show any specific changes and therefore are no indication of oocyte maturation.

Nevertheless, the response of the cytoplasm to the eventually penetrating sperm shows evidence of cytoplasmic maturation. When oocytes are inseminated for *in vitro* fertilization, some immature oocytes in the germinal vesicle stage can be found in the oocyte cohort: when a sperm eventually penetrates them, it decondenses very little. Decondensation increases parallel to oocyte maturation. In Metaphase II (M II) oocytes, chromosomes and paternal prematurely condensed chromosomes are observed beside the first polar body chromosomes. Only when oocytes which are fully mature, cytoplasmically, are fertilized (later than 34 h post–hCG administration), is a normal male pronucleus observed beside the female pronucleus. Consequently, cytoplasmic maturation is completed later than nuclear maturation. Kubiak[9], in the mouse studies, and Van Wissen and colleagues[10], in human, described the chronology of this cytoplasmic maturation most precisely.

The striking obvious change observed in this period is the progress of meiosis: at 20 h post–hCG administration the nucleus is in Metaphase I and progresses to M II at 32 h, when cytoplasmic maturation is not yet achieved. After 32 h the cumulus continues to dissociate (Figure 4); until ovulation after 40 h the follicle continues its growth, to a size of 25–30 mm diam. This size is probably a necessity for ovulation and is much larger than the one necessary to respond to the LH surge.

In the spontaneous cycle, both the dominant follicle (or two follicles in the case of double ovulation) and the small cohort follicles remain healthy until ovulation, despite a rise in the androgen rate in the follicular fluid, but are able neither to achieve ovulatory size nor to resume meiosis[11]. After clomiphene citrate/hMG stimulation, followed by an LH surge, only the largest follicles show mature oocytes; medium follicles (< 16 mm) contain oocytes in germinal vesicles. Human chorionic gonadotrophin is a potent inducer of meiosis, possibly from the effect of the bolus injection; both large and most medium follicles resume meiosis.

The success of GnRH analogue treatment, which produces a large number of oocytes in M II from 12 mm to > 20 mm diam., may be related to the prior preparation for resumption of meiosis at time 0, before the LH surge or hCG administration (Figure 5).

SOME ABNORMAL FOLLICLE MORPHOLOGIES

A well-known anomaly of the follicle is the luteinized unruptured follicle (LUF) syndrome, which is the consequence of an insufficient rise of LH or

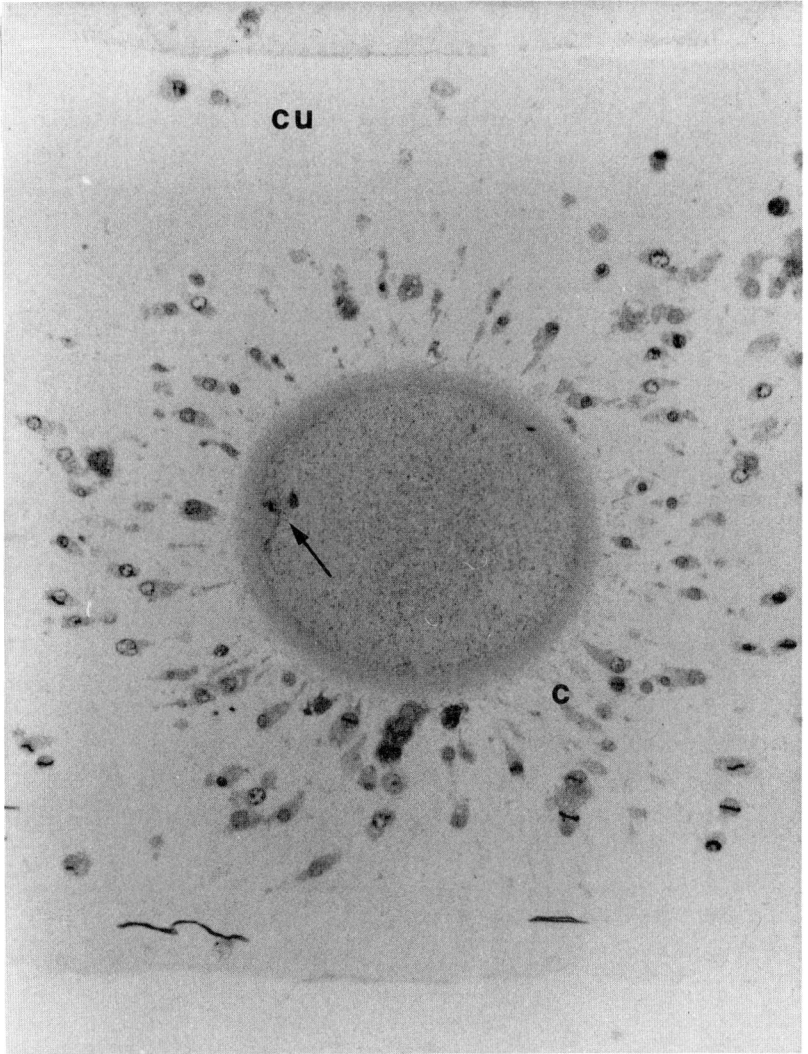

Figure 4 Human oocyte in Metaphase II (M II). The arrow shows M II and the first polar body. cu, dissociated cumulus; c, elongated corona. (× 330)

an insufficient hCG administration. Luteinization of the granulosa corresponds to the progressive increase of progesterone production during the last stage of preovulatory maturation, which continues after ovulation.

22

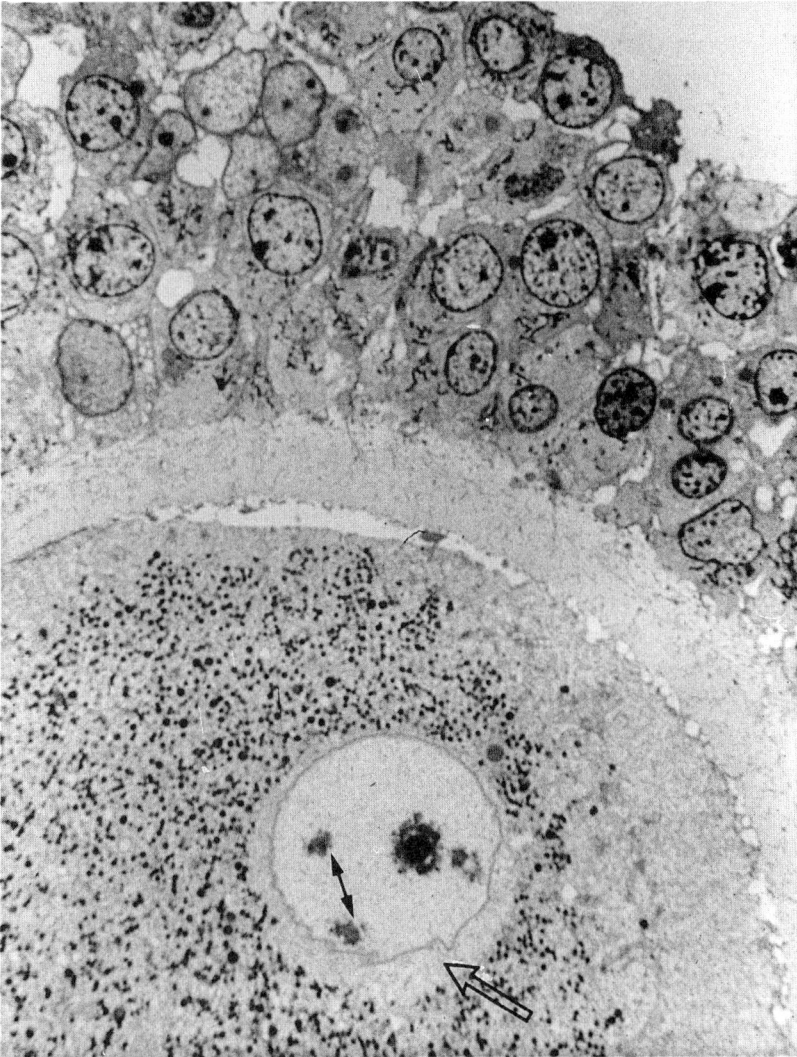

Figure 5 D-TRP[6] + FSH stimulation: human oocyte at time 0 showing a little dissociation of cumulus and corona. The large arrow shows where the white nuclear membrane is invaginated; small arrows indicate condensing black chromosomes. (x 825)

With an insufficient rise of LH, the transformation of the follicle which allows the expulsion of the oocyte does not occur; in particular, the desegregation of the follicle near the apex and blood stasis around the follicle does not occur. The suprafollicular epithelium keeps its cuboidal aspect, and rupture of the follicles does not take place. The luteinization of the cumulus corona complex, which was not preluteinized as the granulosa cells, despite their production of progesterone, occurs very quickly after the probable time for ovulation. Luteinization occurs progressively from the oocyte to the border of the cumulus. A few hours after the normal time–limit for ovulation (i.e. 40 h after hCG administration), the cumulus is still dissociated, but the oocyte is already surrounded by a ring of luteinized cells, which is probably an obstacle to the penetrating sperm if overmature oocytes are retrieved for IVF purposes[12].

Another follicular anomaly is the so-called 'empty follicle syndrome'. Although described in several cases recently, it corresponds to an inability for oocyte retrieval. At time 0, before the LH rise, we observed one patient (of 100), in which we were unable to retrieve any oocyte after retrieving and washing, with our usual experimental method, 13 large follicles (Vu N. Huyen and O. Bomsel-Helmreich, personal communication). The histological image of these preovulatory follicles showed normal healthy oocytes inside the follicle but the equally healthy granulosa, which was very convoluted and abnormally compact, impeded the aspiration of the cumulus corona oocyte complex.

CONCLUSION

In the literature, spontaneously maturing oocytes have very rarely been described. In our study of 150 human preovulatory oocytes, only five were obtained in spontaneous cycles, and all the others after diverse stimulation methods. Therefore, even if the general features of development of the follicle and maturation of the oocyte are no different after stimulation protocols (apart from some we have observed such as early preluteinization of the granulosa, resumption of meiosis in medium follicles or premature preparation for meiosis) there are probably other differences which are worth investigating.

Nevertheless, the main chronological events remain the same for all preovulatory follicles: at 15 h post-hCG administration: germinal vesicle breakdown takes place; at 20 h Metaphase I appears: at 32 h M II and polar body chromosomes appear: and at 38 h the oocyte is ready for fertilization.

Two events of the preovulatory phase are often underestimated, but nevertheless important. First, the dissociation of the cumulus is already very advanced at 20 h after hCG administration, when the oocyte is in M I and ovulation is yet to occur 18 h later, so that cumulus observation does not permit the M I oocyte to be distinguished from the M II. Second, there is a long (12-h) delay between M I and M II during which the oocyte is in a most sensitive state, and prone to the induction of possible anomalies of the oocyte.

Clearly, ethical problems interfere with research on human reproduction processes, especially where oocytes and their maturation are concerned. This encourages us to extrapolate from ethically permitted studies, even if we know that their results do not necessarily correspond to the situation about which we wish to know more.

REFERENCES

1. Di Zerega, G.S. and Hodgen, G.D. (1981). Folliculogenesis in the primate ovarian cycle. *Endocr. Rev.*, **2**, 27–54
2. Bomsel-Helmreich, O. (1985). Ultrasound and the preovulatory human follicle. In Clarke, J. (ed.) *Oxford Reviews of Reproductive Biology*, vol. 7, pp. 1–98 (New York: Oxford University Press)
3. Boklage, C.E. (1990). Survival probability of human conceptions from fertilization to term. *Int. J. Fertil.*, **35**, 75–84
4. Al-Mufti, W., Bomsel-Helmreich, O. and Christides, J. P. (1988). Oocyte size and intrafollicular position in polyovular follicles in rabbits. *J. Reprod. Fert.*, **82**, 15–25
5. Bomsel-Helmreich, O., Durand-Gasselin, I., Pennehouat, G., Vu N. Huyen, L., Antoine, J.M. and Salat-Baroux, J. (1990). Structure and ultrastructural of human preovulatory oocytes immediately before the administration of hCG. In Evers, J.H.L. and Heinemann, M.J. (eds.) *From Ovum to Implantation*, pp. 81–90 (Amsterdam: Elsevier)
6. Bomsel-Helmreich, O., Vu N. Huyen, L., Durand-Gasselin, I., Salat-Baroux, J. and Antoine, J.M. (1987). Timing of nuclear maturation and cumulus dissociation in human oocytes stimulated by CC, hMG and hCG. *Fertil. Steril.*, **48**, 586–95
7. Espey, L.L. (1974). Ovarian proteolytic enzymes and ovulation. *Biol. Reprod.*, **10**, 216–35
8. Al-Mufti, W., Bomsel-Helmreich, O. and Christides, J.P. (1993).

Maturational changes of the human zona pellucida during the preovulatory period. *Hum. Reprod.*, in press

9. Kubiak, J.Z. (1989). Mouse oocytes gradually develop the capacity for activation during the metaphase II arrest. *Developmental Biology*, **136**, 537–45

10. Van Wissen, B., Bomsel-Helmreich, O., Debey, P., Eisenberg, C., Vautier, D. and Pennehouat, G. (1991). Fertilization and ageing processes in non-divided human oocytes after GnRHa treatment: an analysis of individual oocytes. *Hum. Reprod.*, **6**, 879–84

11. Bomsel-Helmreich, O., Gougeon, A., Thebault, A., Saltarelli, D., Milgrom, E., Frydman, R. and Papiernik, E. (1979). Healthy and atretic human follicles in the preovulatory phase: differences in the evolution of follicular morphology and steroid content in the follicular fluid. *J. Clin. Endocrinol. Metab.*, **48**, 686–94

12. Bomsel-Helmreich, O., Vu N. Huyen, L., Durand-Gasselin, I., Salat-Baroux, J. and Antoine, J.M. (1987). Mature and immature oocytes in large and medium follicles after CC and hMG stimulation without hCG. *Fertil. Steril.*, **48**, 596–604

3

Local control of ovarian function

S. G. Hillier

INTRODUCTION

Oestradiol secreted by the preovulatory follicle triggers the ovulation-inducing luteinizing hormone (LH) surge and prepares the female reproductive tract for conception. Knowledge of the endocrine and paracrine mechanisms that regulate follicular oestrogen synthesis is therefore vital to reproductive medicine. Here, our current understanding of these mechanisms is summarized, specifying the individual functions of the gonadotrophins follicle-stimulating hormone (FSH) and LH and highlighting the local interactions between granulosa and thecal cells that are crucial to ovarian oestrogen synthesis [1,2].

THE TWO-CELL, TWO-GONADOTROPHIN HYPOTHESIS

The requirement for *both* FSH and LH to stimulate normal follicular development and oestrogen synthesis was recognized half a century ago. Fevold [3] treated immature or hypophysectomized rats with highly purified preparations of ovine pituitary FSH and LH, measuring ovarian and

This article has previously been published in the *Proceedings of the Ninth International Congress of Endocrinology*, Nice, 1992 and is reproduced here with the permission of the editors uterine weights to demonstrate that FSH alone was relatively inefficient in inducing the ovaries to secrete oestrogen, and that whereas LH alone was unable to stimulate oestrogen secretion, it augmented secretion when co-injected with FSH. Greep and colleagues[4] extended these findings, providing the first histological evidence that FSH acts on the granulosa cell layer (induction of follicular antrum formation) and LH on the theca (induction of thecal/interstitial hypertrophy).

Direct evidence that both granulosa and thecal cells are involved in the follicular synthesis of oestrogen was provided by the classical microdissection/autotransplantation studies of Falck[5]. He separated granulosa cells and thecal cells from rat follicles and transplanted each cell type alone or in combination to the anterior chamber of the eyes of ovariectomized rats, a contiguous vaginal autotransplant serving as an indicator of oestrogen production by the ovarian grafts. Results showed that when granulosa cells and thecal interstitial cells were transplanted alone, neither cell type alone induced cornification of the vaginal autotransplant. However, transplants containing thecal/interstitial cells combined with granulosa cells consistently induced vaginal cornification, leading him to conclude 'that the production of oestrogen is dependent on an interplay between theca interna gland cells or interstitial cells ... and granulosa cells'.

The use of radioisotopically labelled precursors to elucidate biochemical pathways of ovarian steroid production and metabolism in the 1960s provided biochemical evidence of the importance of granulosa–theca interaction to follicular oestrogen synthesis. Ryan and colleagues[6] showed that separated theca interna and granulosa cell preparations from preovulatory follicles could each synthesize oestrogen from acetate, but that the yield of labelled oestrogen was more than additive when the two cell types were incubated together. It therefore followed that both granulosa and thecal cells contained the enzyme activity crucial to oestrogen synthesis, aromatase.

The ability of (porcine) granulosa cells to undertake extensive metabolism of androgen to oestrogen was confirmed by Bjersing[7]. However, he also noted that granulosa cells did not bear the morphological hallmarks of steroid-secreting cells (lipid droplets, extensive endoplasmic reticulum, etc.) and were relatively deficient in 17-hydroxylase and C_{17-20} lyase, enzymic activities required to synthesize the C_{19} androgens (androstene-dione and testosterone) required for aromatization to oestrogens (oestrone and oestradiol)[7,8]. On the other hand, thecal cells had both the morphological and biochemical characteristics of a steroidogenic cell type, being able to synthesize both androgens and oestrogens from acetate or C_{21} steroid precursors *in vitro*[6,7]. Hence the first 'two-cell type' theory of oestrogen synthesis, put forward by Short, regarded the (equine) theca interna as the major site of follicular oestrogen formation from granulosa-derived progesterone[8].

The view that thecal cells were primary cellular sites of follicular oestrogen formation held sway until the end of the 1970s[9], when it was revised in the light of experiments using highly sensitive and specific steroid radioimmunoassays to monitor steroid production by isolated follicular cell types cultured *in vitro*. Such studies of granulosa cells isolated from rat ovaries showed that despite the inability of these cells to synthesize androgens *de novo*, they readily aromatized exogenous androgen to oestradiol when stimulated with FSH[10]. On the other hand, thecal/interstitial cells were deficient in aromatase activity but produced increased amounts of androgen in response to stimulation by LH[11]. Follicular oestrogen synthesis *in vivo* could also be induced by treating hypophysectomized animals with FSH plus an aromatizable androgen such as testosterone, bypassing the need for LH to stimulate endogenous androgen synthesis[12]. Granulosa cells were also shown to be the exclusive sites of FSH receptor expression in the ovaries, whereas LH receptors were detected on thecal/interstitial cells and only appeared on granulosa cells during advanced stages of preovulatory development stimulated by FSH (reviewed elsewhere[13]). Hence the formulation of Armstrong and Dorrington's 'two-cell, two-gonadotrophin' mechanism[14] in which the theca interna came to be regarded as the primary intrafollicular site of androgen synthesis regulated by LH, and the granulosa cell layer the major site of oestrogen formation (aromatization) stimulated by FSH.

The relevance of this mechanism to follicular oestrogen synthesis in the human ovary was established using similar experimental techniques (reviewed elsewhere[15,16]). Comparisons of aromatase activity in granulosa and thecal cells isolated from human preovulatory follicles showed that more than 99.9% of the total follicular activity resided in the granulosa cell layer. More recent experiments based on molecular approaches have confirmed that both aromatase cytochrome P450 mRNA[2] and protein[17] are selectively localized to the granulosa cell layer of the human preovulatory follicle.

ENDOCRINE REGULATION OF FOLLICULAR OESTROGEN SYNTHESIS

The initiation, maintenance and termination of oestrogen secretion by the human preovulatory follicle depend on development-dependent changes in responsiveness to FSH and LH, and can be summarized as follows (see elsewhere for reviews[1,13,16,18,19]).

FSH receptors are located on granulosa cells and LH receptors on thecal cells throughout antral follicular development. Incipient (< 5 mm diameter) preovulatory follicles present at the beginning of a menstrual cycle continue to develop in response to the intercycle rise in plasma FSH levels. By the mid-follicular phase, a single dominant follicle emerges as the largest (≥ 10 mm diameter) healthy follicle in either ovary. Due to stimulation by FSH, granulosa cells in this follicle will have proliferated, acquiring LH receptors and aromatase cytochrome P450[20,21]. Increased expression of aromatase is a hallmark of granulosa cell differentiation and explains the role of these cells as exclusive sites of oestrogen synthesis in the preovulatory follicle. During the mid to late follicular phase, as pre-ovulatory follicular growth continues and oestrogen synthesis increases, granulosa cells become increasingly responsive to FSH and LH[20]. Since the LH receptors on granulosa cells are functionally coupled to aromatase, LH can directly regulate androgen synthesis (in thecal cells), aromatization (in granulosa cells) and hence oestrogen synthesis in the preovulatory follicle. Ovulation and termination of follicular oestrogen synthesis is then induced by the mid-cycle LH surge, triggered by oestradiol once the follicle has fully matured and its oestrogen secretion rate is maximal.

LH-regulated androgen synthesis occurs in thecal cells throughout antral follicular development. LH receptors are constitutively present on these cells, functionally linked to P450c17 (the steroidogenic cytochrome P450 with both 17-hydroxylase and C_{17-20} lyase activities) via cyclic adenosine monophosphate (AMP)-mediated post-receptor signalling[22]. Thecal cells also possess insulin-like growth factor (IGF) receptors that are functionally coupled to androgen synthesis[2,23]. Presence of IGF-I in human thecal cell culture medium therefore promotes LH/cyclic AMP-stimulated androgen synthesis *in vitro*, consistent with a 'co-gonadotrophic' function for IGFs of hepatic (endocrine) or granulosa (paracrine) origin in modulating follicular androgen synthesis *in vivo* (reviewed elsewhere[2]).

PARACRINE REGULATION OF FOLLICULAR OESTROGEN SYNTHESIS

The 'two-cell, two-gonadotrophin' model also provides a conceptual framework for understanding paracrine signalling within the follicle wall. Paracrine communication between granulosa and thecal cells in developing follicles has been likened to a generalized epithelio-mesenchymal interaction in which the epithelial (granulosa) component responds selectively to endocrine stimulation by FSH and the mesenchymal (thecal) component responds to LH[16], both cell types producing steroidal and non-steroidal regulatory factors and an extracellular matrix capable of influencing cell proliferation, differentiation and migration in the maturing follicle[16,24,25].

Theca-derived paracrine control

The theca interna produces diverse steroidal and non-steroidal factors with potentials to influence granulosa cell function *in vivo*[16,24,25]. These include the androgenic steroids produced in response to stimulation by LH. Not only are they aromatase substrates (Δ^4 3-oxoandrogens) and potential competitive aromatase inhibitors (5α-reduced androgens), but they can also promote granulosa cell differentiation in response to FSH *in vitro*, including enhanced induction of aromatase. This regulatory action of androgen is mediated by granulosa-cell androgen receptors and attendant amplification of cyclic AMP-mediated intracellular signalling (reviewed elsewhere[15,26]).

Granulosa-derived paracrine control

Production of inhibin is conspicuous among the many FSH-inducible granulosa cell functions enhanced by androgens *in vitro*[27]. Inhibin has been shown to promote LH-dependent thecal androgen synthesis *in vitro*[28,29]. Thus, the potential exists for a reciprocal interaction between granulosa-derived inhibin and theca-derived androgen which may help explain the development-related increases in oestradiol synthesis that are known to occur in the preovulatory follicle[2].

31

In vivo evidence for paracrine control

The close quantitative relationship that exists between the supply of aromatase substrate by the theca and the rate of aromatization in the granulosa of the human preovulatory follicle has prompted the suggestion 'that a local feedback interaction (granulosa on theca) may be involved in the maintenance of intrafollicular androgen levels'[30]. Molecularly pure, recombinant FSH has now been used to verify this hypothesis directly[31,32]. FSH preparations purified from natural sources (pituitary or urine) are usually contaminated with trace amounts of LH sufficient to stimulate thecal cell function directly when injected into experimental animals, precluding assessment of putative paracrine effects secondary to the action of FSH. The recent availability of recombinant human FSH (rh-FSH), completely devoid of LH, therefore affords a unique opportunity to determine whether granulosa-derived paracrine signalling is promoted by FSH *in vivo*. Preliminary results of injecting rh-FSH into intact, immature female rats have revealed a dose-dependent stimulation of ovarian cytochrome P450c17 mRNA levels[31] and an augmented LH-responsive androgen synthesis *in vitro* by thecal/interstitial cells isolated from rh-FSH-treated animals[32]. Since cytochrome P450c17 expression in the ovary is predominantly located in thecal/interstitial cells (which do not express FSH receptors) these data constitute the firmest evidence to date that intrafollicular paracrine signalling relevant to oestrogen synthesis does indeed occur *in vivo*.

REFERENCES

1. Hillier, S.G. (1990). Ovarian manipulation with pure gonadotrophins. *J. Endocrinol.*, **127**, 1–4
2. Hillier, S.G. (1991). Paracrine control of follicular oestrogen synthesis. *Semin. Reprod. Endocrinol.*, **9**, 332–40
3. Fevold, H.L. (1941). Synergism of the follicle stimulating and luteinizing hormones in producing estrogen secretion. *Endocrinology*, **28**, 33–6
4. Greep, R.O., Van Dyke, H.B. and Chow, B. (1942). Gonadotropins of the swine pituitary 1. Various biological effects of purified thylakentrin (FSH) and pure metakentrin (ICSH). *Endocrinology*, **30**, 635–49
5. Falck, B. (1959). Site of production of oestrogen in rat ovary as studied by microtransplants. *Acta Physiol. Scand.*, **47**, Suppl. 163, 1–101

6. Ryan, K.J., Petro, Z. and Kaiser, J. (1968). Steroid formation by isolated and recombined ovarian granulosa and theca cells. *J. Clin. Endocrinol. Metab.*, **28**, 355–8

7. Bjersing, L. (1967). On the morphology and endocrine function of granulosa cells in ovarian follicles and corpora lutea: biochemical, histochemical and ultrastructural studies on the porcine ovary with special reference to steroid hormone synthesis. *Acta Endocrinol.*, Suppl. **125**

8. Short, R.V. (1962). Steroids in the follicular fluid and corpus luteum of the mare. A 'two-cell type' theory of ovarian steroid synthesis. *J. Endocrinol.*, **24**, 59–63

9. Ryan, K.J. (1979). Granulosa-thecal cell interaction in ovarian steroidogenesis. *J. Steroid Biochem.*, **11**, 799–800

10. Dorrington, J.H. and Armstrong, D.T. (1975). Estradiol-17β synthesis in cultured granulosa cells from hypophysectomized immature rats: stimulation by follicle-stimulating hormone. *Endocrinology*, **97**, 1328–31

11. Fortune, J.E. and Armstrong, D.T. (1977). Androgen production by theca and granulosa cells isolated from proestrous rat follicles. *Endocrinology*, **100**, 1341–7

12. Armstrong, D.T. and Papkoff, H. (1976). Stimulation of aromatization of exogenous androgens in ovaries of hypophysectomized rats *in vivo* by follicle-stimulating hormone. *Endocrinology*, **99**, 1144–51

13. Zeleznik, A.J. and Hillier, S.G. (1984). The role of gonadotropins in the selection of the preovulatory follicle. *Clin. Obstet. Gynecol.*, **27**, 927–40

14. Armstrong, D.T. and Dorrington, J.H. (1979). Estrogen biosynthesis in the ovaries and testes. In Thomas, J.A. and Singhal, R.L. (eds.) *Regulatory Mechanisms Affecting Gonadal Hormone Action*, Vol. 2, pp. 217–58. (Baltimore: University Park Press)

15. Hillier, S.G. (1985). Sex steroid metabolism and follicular development in the ovaries. In Clarke, J.R. (ed.) *Oxford Reviews of Reproductive Biology*, Vol. 7, pp. 168–222. (Oxford: Clarendon Press)

16. Hillier, S.G. (1991). Cellular basis of follicular endocrine function. In Hillier, S.G. (ed.) *Ovarian Endocrinology*, pp. 73–106. (Oxford: Blackwell Scientific Publications)

17. Sasano, H., Okamoto, M., Mason, J.I. *et al.* (1989). Immunolocalization of aromatase, 17β-hydroxylase and side-chain cleavage cytochromes P-450 in the human ovary. *J. Reprod. Fertil.*, **85**, 163–9

18. Hsueh, A.J.W., Adashi, E.Y., Jones, P.B.C. and Welsh, T.J. Jr (1984). Hormonal regulation of the differentiation of cultured granulosa cells. *Endocr. Rev.*, **5**, 76–126

19. Richards, J.S., Jahnsen, T., Hedin, L. *et al.* (1987). Ovarian follicular development: from physiology to molecular biology. *Recent Prog. Horm. Res.*, **43**, 231–70

20. Zeleznik, A.J. and Kubik, C.J. (1986). Ovarian responses in macaques to pulsatile infusion of follicle-stimulating hormone (FSH) and luteinizing hormone (LH): increased sensitivity of the maturing follicle to FSH. *Endocrinology*, **119**, 2025–32

21. Hickey, G.J., Chen, S., Besman, M.J. *et al.* (1988). Hormonal regulation, tissue distribution, and content of aromatase cytochrome P450 messenger ribonucleic acid and enzyme in rat ovarian follicles and corpora lutea: relationship to estradiol biosynthesis. *Endocrinology*, **122**, 1426–36

22. Erickson, G.F., Magoffin, D.A., Dyer, C.A. and Hofeditz, C. (1985). The ovarian androgen producing cells: a review of structure/function relation-ships. *Endocr. Rev.*, **6**, 371–99

23. Bergh, C., Olsson, J.H., Carlsson, B. *et al.* (1992). Regulation of androgen biosynthesis in cultured human thecal cells by LH, IGF-I and insulin. Presented at the *IX Ovarian Workshop Ovarian Cell Interaction: Genes to Physiology*, July, Chapel Hill (poster abstract 22)

24. Tonetta, S.T. and DiZerega, G.S. (1989). Intragonadal regulation of follicular maturation. *Endocr. Rev.*, **10**, 205–29

25. Adashi, E.Y. (1989). Putative intraovarian regulators. *Semin. Reprod. Endocrinol.*, **7**, 1–100

26. Daniel, S.A.J. and Armstrong, D.T.(1986). Androgens in the ovarian microenvironment. *Semin. Reprod. Endocrinol.*, **4**, 89–100

27. Hillier, S.G., Wickings, E.J., Illingworth, P.I. *et al.* (1991). Control of immunoactive inhibin production by human granulosa cells. *Clin. Endocrinol.*, **35**, 71–8

28. Hsueh, A.J.W., Dahl, K.D., Vaughan, J. *et al.* (1987). Heterodimers and homodimers of inhibin subunits have different paracrine action in the modulation of luteinizing hormone-stimulated androgen biosynthesis. *Proc. Natl. Acad. Sci. USA*, **84**, 5082–6

29. Hillier, S.G., Yong, E.L., Illingworth, P.I. *et al.* (1991). Effect of recombinant inhibin on androgen synthesis in cultured human thecal cells. *Mol. Cell. Endocrinol.*, **75**, R1–6; **79**, 177

30. Hillier, S.G., Reichert, L.E. Jr and van Hall, E.V. (1981). Control of preovulatory follicular estrogen biosynthesis in the human ovary. *J. Clin. Endocrinol. Metab.*, **52**, 847–56

31. Smyth, C.D., Whitelaw, P.F., Turner, I.M. *et al.* (1992). Paracrine control of ovarian endocrine function. In Hillier, S.G.(ed.) *Gonadal Development and Function*, pp. 145–8. (New York: Raven Press)

32. Smyth, C.D., Miró, F., Howles, C.M. and Hillier, S.G. (1992). Modulation by inhibin and insulin-like growth factor I (IGF-I) of LH-responsive androgen synthesis in rat ovarian theca/interstitial cell cultures. *J. Endocrinol.*, **132** (Suppl.) Abstr. 46

4

Follicular maturation in childhood and puberty

I.A. Hughes

INTRODUCTION

The integration and synchronization of hormonal events which lead to ovulation is the end result of a maturational process starting in early fetal life. The ovary is far from being a quiescent endocrine gland before puberty, yet its development has traditionally been described as resulting from the absence of a testis-determining factor (now known as the SRY gene[1]) rather than from the presence of any hitherto unrecognized factors located on the X chromosome. There is clinical, hormonal and ultrasonographic evidence to indicate that ovarian folliculogenesis occurs during fetal life, infancy and childhood, as well as at puberty. Furthermore, there are several clinical disorders which serve to illustrate the remarkable sensitivity of target tissues to the low levels of oestrogen produced during infancy and childhood.

EMBRYOLOGY OF THE OVARY

The primordial germ cells of the indifferent gonad originate, in both sexes, from the endoderm of the yolk sac where they migrate by the 6th week of gestation to the genital ridge on the medial surface of the mesonephros. Contact between primitive gonadal and renal tissues is maintained in the male in the form of a rete testis, whereas in the female this connection disappears. The unique feature of ovarian development is the rapid multiplication of germ cells (oogonia) to reach a peak number of approximately 7 million by 24 weeks of gestation (Figure 1). Many oogonia reach and remain suspended in the prophase stage of the meiotic division and are then termed oocytes. The first meiotic division is not completed until ovulation (approximately 14 years later), and the second not until fertilization when the ovum is restored with a haploid number of chromosomes. The atretic process accelerates rapidly from fetal to postnatal life so that by puberty the number of oocytes is reduced to 400 000. Atresia is even more accentuated in Turner's syndrome (gonadal dysgenesis) when streak ovaries are usually evident by early childhood.

Follicular maturation starts when the primary oocyte becomes surrounded by a single layer of granulosa cells to form a primordial follicle. When several layers of cuboidal-shaped granulosa cells have formed around the oocyte, a primary follicle is formed. The stromal cells adjacent to the basal lamina membrane differentiate to form thecal cells. Thus, the two-cell format of the ovary (granulosa and thecal cells) is established early although their independent and complementary steroid responsiveness to the two gonadotrophins, follicle stimulating hormone (FSH) and luteinizing hormone (LH), is not established until later in development. Further maturation of the primary follicle to form an antral follicle takes place after puberty and through the reproductive period. Until this stage the process is gonadotrophin-independent but FSH then plays a role in 'rescuing' small antral follicles from atresia.

PHYSICAL SIGNS OF FOLLICULAR MATURATION

Some of the physical consequences of folliculogenesis in childhood are listed in Table 1. Breast enlargement is common in both sexes at birth due to the effect of maternal oestrogens. There may even be some secretion from the nipple, the so-called 'witch's milk'. The infant vagina may also respond to oestrogen action in the form of a white discharge with occasional spotting of blood (a withdrawal bleed). Evidence based on ultrasonography indicates that the uterus is also larger at birth because of maternal oestrogenic stimulation. Thereafter the uterine corpus decreases in size so that the cervix is the predominant component of the uterus. The cervix is easily palpable on rectal examination.

6 weeks	Primordial germ cells	
	migrate	
	Genital ridge	
12–24 weeks	Oogonia proliferation	7 000 000
	Primary oocytes in meiotic arrest	
Birth	2 000 000 oocytes	A T R E S I A
Puberty	400 000 oocytes	
Reproductive span	400 oocytes	

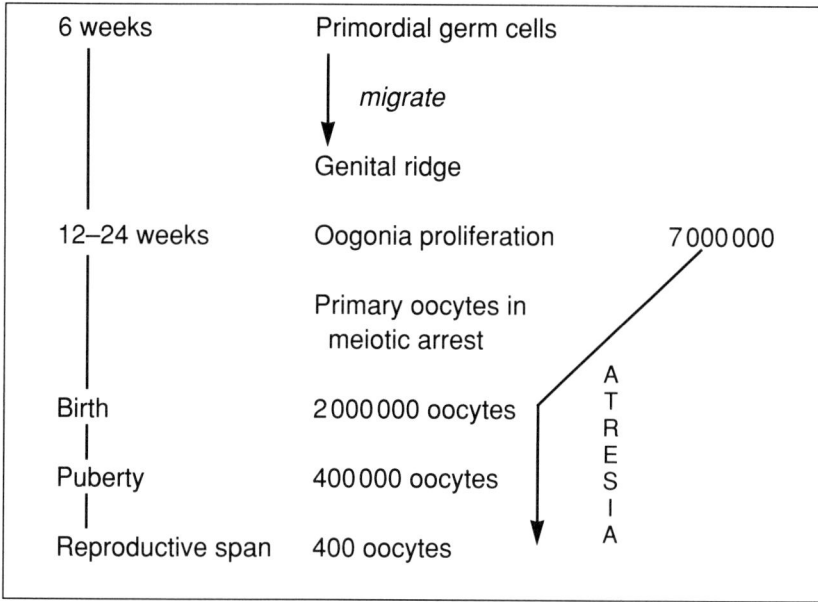

Figure 1 Schema depicting the chronology of ovarian development

No further signs of oestrogen action occur until the time of puberty. Breast development starts on average at 11 years of age, although the range of normality extends from 8 to 13 years. Five stages are described, Stage 1 being the prepubertal state and Stage 5 the mature adult breast[2]. The onset of breast development (Stage 2) is heralded by a 'bud' formation from both breast and papilla elevation and an increase in size of the areola. The swelling is firm on palpation and can typically be distinguished from soft adipose tissue because the underlying rib cannot be felt. The remaining stages represent progressive changes in breast size and shape; some girls remain at Stage 4 until well after the completion of puberty but the maturational process from Stages 2 to 5 generally takes 4 years. Breast development is the first of the secondary sexual characteristics to appear, closely followed by pubic and labial hair growth. The temporal relationship of these events to each other is illustrated in Figure 2. The growth spurt, mediated by oestrogens together with an increase in growth hormone secretion, occurs relatively early in the pubertal process. Peak

Table 1   Physical consequences of folliculogenesis

Breast	Uterus
Onset at 11 years (range 8–13)	Larger size at birth (maternal
Proceeds through five defined stages	oestrogen)
Duration of development 4 years	Corpus lengthening at puberty
	Adult corpus to cervix ratio
	Cervical mucoid secretion
Vagina/vulva	
Layers of superficial cells develop	
Acid vaginal fluid (lactogenic	Anthropometric
bacilli)	Growth spurt
Vaginal lengthening and rugae	Fat deposition
Increase in fat on mons pubis and	Broadening of the pelvis
labia majora	(gynaecoid)
Swelling of labia minora	Skeletal maturation
Hymen thickening	

height velocity is achieved at around breast Stage 3. Of particular note is the timing of menarche, occurring after the peak growth spurt. The average age of menarche among Europeans is currently 12.8 years with a wide range of variation. In Britain, mean ages of menarche in the south and north are 13.0 and 13.3 years, respectively[3]. Statural growth ceases when skeletal epiphyses fuse, usually corresponding to a chronological age of 16–17 years.

Signs of oestrogenization are evident in vaginal lengthening and in its lining mucosa. Before puberty the mucosa is composed of a thin layer of epithelial cells and the vaginal secretions are alkaline. The vaginal mucosa becomes stratified into layers of squamous epithelium and the pH changes to acid. The hymen thickens and its orifice enlarges. There is increased fat deposition over the mons pubis and the labia increase in size. The uterine corpus : cervix ratio throughout childhood is 1 : 2; first, there is corpus lengthening followed by an increase in the thickness of both uterine components. The cervical glands produce a mucoid secretion which, when left to dry in thin preparations, produces a typical fern-like crystal appearance indicative of oestrogen stimulation.

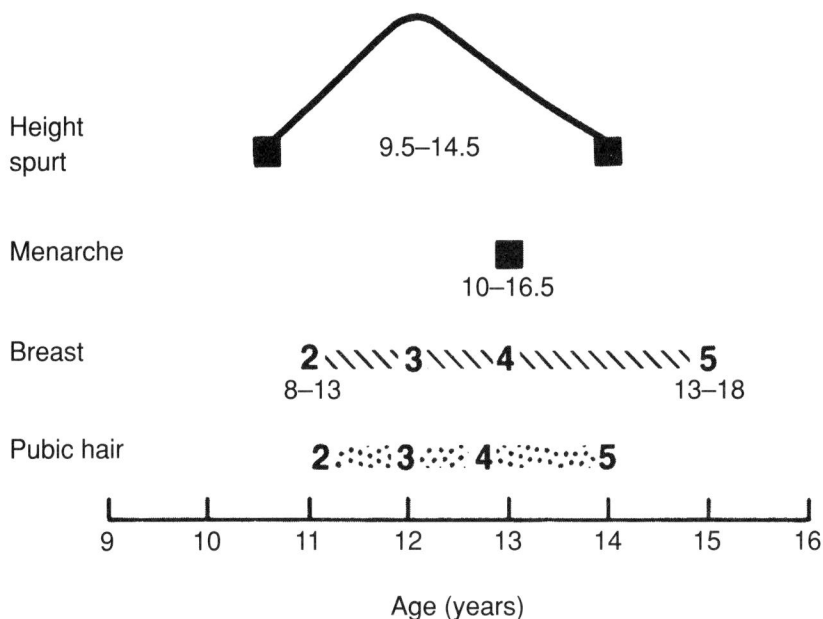

Figure 2 Temporal relationship of the signs of puberty in the female. Age-ranges are given for pubertal events. The Tanner staging for puberty is reflected by the numbers 2–5. Reproduced with permission from ref. 11

ENDOCRINOLOGY OF FOLLICULAR MATURATION

The developing ovary is exposed to fluctuating concentrations of gonadotrophins during three phases of life before adult maturity is reached. Initially in fetal life, there is an increase in placental human chorionic gonadotrophin (hCG) secretion which reaches its peak at 10 weeks of gestation. Concomitant with this rise, there is also an increase in fetal LH and FSH secretion which reaches a maximum at mid-gestation. There is a slight increase in fetal serum oestradiol concentrations, but the rise is not as profound as for testosterone in the male fetus which is essential for normal male genital development. There is also evidence from studies of fetal steroidogenesis, using measurement of specific mRNA synthesis, that the maturation of ovarian aromatase activity is delayed, relative to other steroidogenic enzymes[4]. Fetal gonadotrophin concentrations increase gradually towards term. Cord blood levels of oestradiol are very high but fall rapidly after birth. There is a postnatal surge in FSH and LH levels in both sexes. In the female, the rise in FSH is greater and more prolonged than it is in the male. This produces a slight rise in plasma oestradiol concentrations. In contrast, the rise in LH levels in the male infant leads to a pronounced increase in plasma testosterone concentrations which may approach the lower end of the normal adult male range.

There is little endocrine evidence of ovarian activity during childhood, in contrast to studies using ultrasound. Plasma gonadotrophin and oestradiol levels are low; the latter remaining below the detection limit of most immunoassays for oestradiol. FSH levels generally remain higher than LH throughout childhood. The onset of puberty is accompanied by an increase in gonadotrophin production and ovarian steroid secretion. A detailed description of the neuroendocrine events which trigger puberty is beyond the scope of this article[5]. Figure 3 summarizes the principal factors associated with the onset of puberty in females. Non-migration of the gonadotrophin releasing hormone (GnRH) neurons from the olfactory placode, for example, is a cause of the Kallmann's syndrome. A central role is played by the hypothalamic GnRH pulse generator which is responsible for initiating increased nocturnal pulses of LH secretion as one of the first endocrine signs of puberty. The responsiveness of the pituitary to acute stimulation by GnRH is also enhanced before any physical signs of puberty develop. An increase in pulses of LH and FSH secretion both day and night is followed subsequently by a cyclical pattern of gonadotrophin secretion characteristic of the mature female. Throughout this period, there will have been a gradual increase in ovarian secretion of steroids such as oestradiol, 17OH-progesterone, progesterone and androstenedione. Which factor(s) actually activates the GnRH pulse generator at the expected time of puberty is still unknown. Extrinsic factors such as nutrition, as well as a host of neurotransmitter substances, all play important roles in a complex interdependent manner.

Menarche is a relatively late event in the process of pubertal maturation. Initially the cycles are usually non-ovulatory although ovulation from the outset of menses occurs in some girls. The results of several studies indicate that there is typically an interlude of about 2 years following menarche before ovulatory cycles appear. Figure 4 shows the results of one such study which used profiles of salivary progesterone concentrations as an index of ovulation[6]. In a group of healthy girls it was possible to

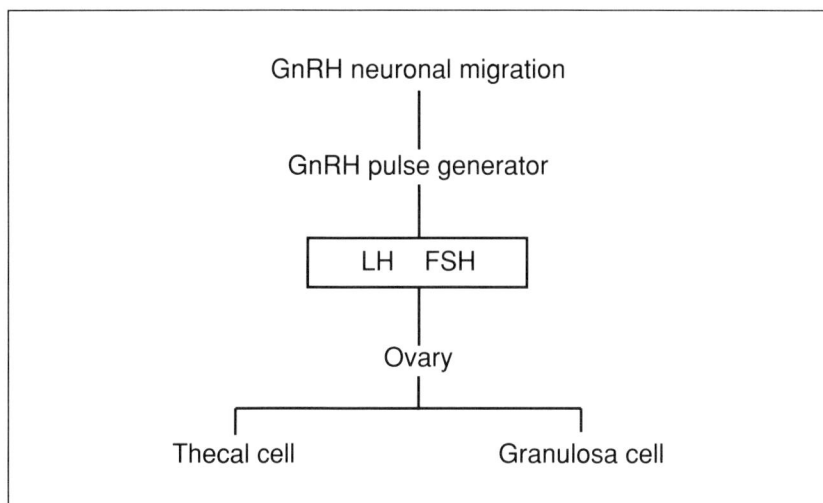

Figure 3 Factors associated with the control of the onset of female puberty

identify five patterns or grades of progesterone profiles. Grades 1, 2 and 3 were associated with spikes of salivary progesterone levels no greater than 40, 150 and 300 pmol/l, respectively. By Grade 4, there was a biphasic pattern predominating with a 14-day span of elevated progesterone levels. Grade 5 showed the typical ovulatory pattern of the mature female. The acquisition of a Grade 4 or 5 pattern of progesterone profile was consistent with a probable ovulatory cycle, although there was no confirmatory evidence on ultrasonography in this study. Analysis of the data (Figure 5) showed there were no Grade 4 or 5 profiles in premenarchal girls, whereas 60–70% of girls who were more than two years post-menarche had evidence of ovulatory cycles.

RADIOLOGICAL EVIDENCE OF FOLLICULAR MATURATION

Evidence of cystic changes in prepubertal ovaries is well described from autopsy studies[7]. However, ultrasonography has shown that single and multiple ovarian cyst formation is not uncommon in normal prepubertal girls, including neonates (Figure 6). Sound ultrasound techniques to assess

41

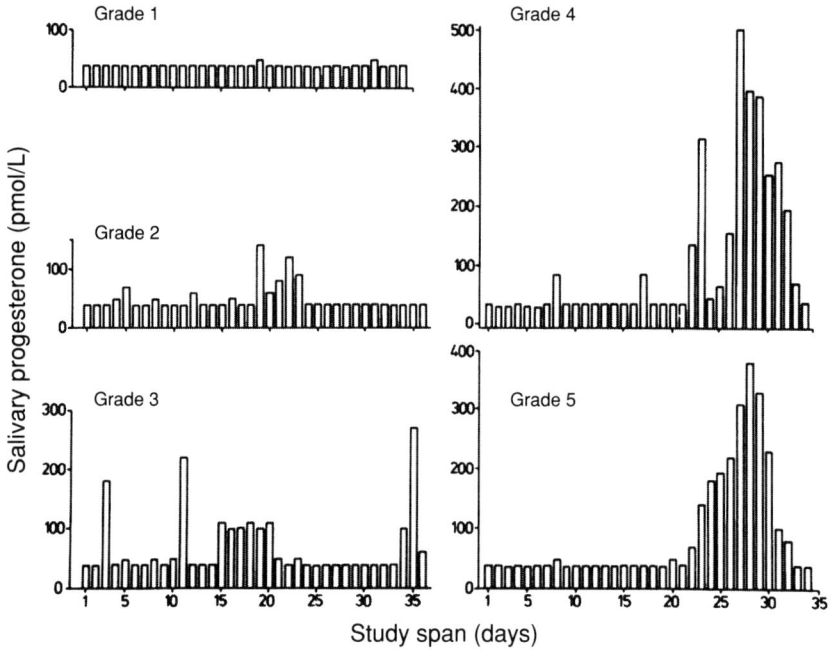

Figure 4 Profiles of progesterone concentrations in saliva graded 1–5 according to increased levels of this steroid during the second half of the menstrual cycle. See text for further descriptive details

ovarian and uterine morphology demand a full bladder before transverse and longitudinal sectional views are taken[8]. The size of the ovary at birth is approximately 15 mm long, 3 mm wide and 2.5 mm thick, from which ovarian volume can be calculated using the appropriate formula for an elliptical solid. Ovarian volume remains fairly constant until about 6 years of age when it gradually increases in size, with an accelerated increase at puberty. By maturity, the ovarian volume ranges from 2 to 6 cm³. During each menstrual cycle a number of primordial follicles start to develop, with a dominant follicle at the time of ovulation measuring between 1.7 and 2.9 cm in diameter.

The uterus is relatively large at birth due to the effect of maternal oestrogens; the mean length is 3.4 cm and a distinct endometrium can be visualized. Uterine size decreases in later infancy and remains approximately 2.5–3.0 cm in length throughout childhood. The shape is

42

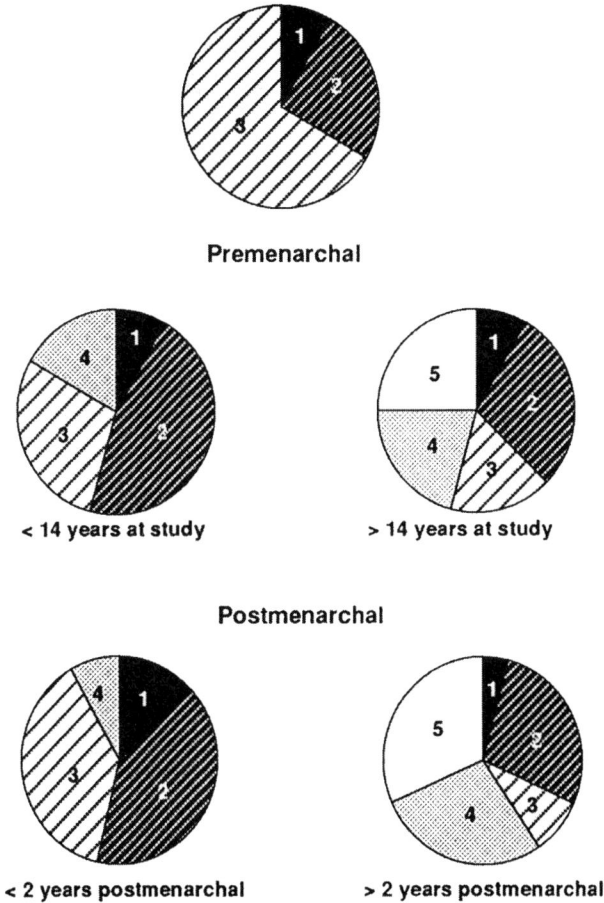

Premenarchal

Postmenarchal

< 14 years at study   > 14 years at study

< 2 years postmenarchal   > 2 years postmenarchal

Figure 5 Analysis of saliva progesterone profile grades (as defined in Figure 4) according to menarchal status in a group of adolescent girls. Postmenarchal girls were subdivided into years since menarche

described as 'tear drop' and the ratio of uterine corpus to cervix is 1 : 2. By postpuberty the ratio is reversed. The prepubertal uterus starts to increase in size from about 7 years of age onwards, so that in the postpubertal girl the length ranges between 5 and 8 cm. The thickness of the endometrium varies throughout the menstual cycle, reaching 5–6 mm in thickness by the mid-luteal phase.

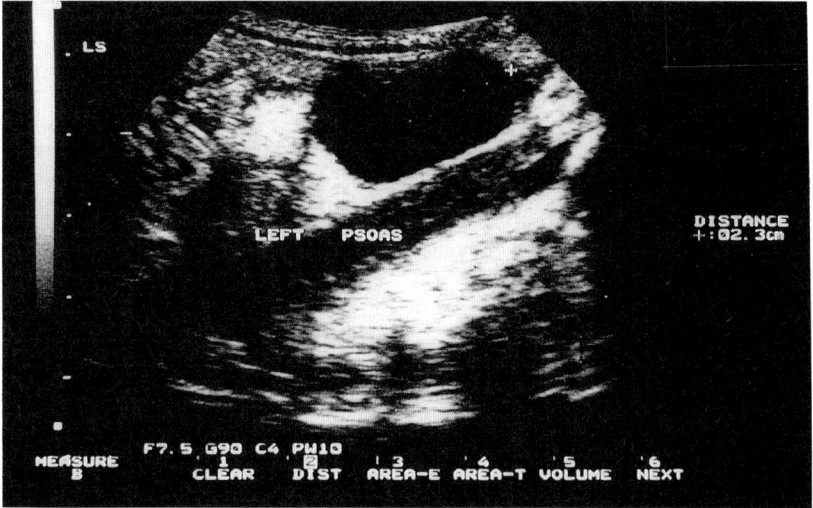

Figure 6 Pelvic ultrasound scan of a female neonate showing large ovarian cyst. (Courtesy of Dr Berman, Department of Radiology, Addenbrooke's Hospital, Cambridge)

CLINICAL DISORDERS

A number of conditions occur in childhood which are associated with either premature activation of ovarian function or delayed onset of the normal pubertal process. These are briefly described merely to illustrate that the effects of folliculogenesis can be apparent before adulthood.

Premature thelarche

This relatively common condition is defined as the development of breast tissue before the age of 8 years in the absence of any other signs of puberty. The typical age of onset is 1–3 years, with the majority presenting before 2 years of age. Breast development may be unilateral or bilateral and characteristically the degree of breast swelling waxes and wanes over time. The rate of linear growth is unchanged and skeletal maturation is not advanced.

The results of endocrine investigations are generally appropriate for the prepubertal age. In particular, plasma oestradiol levels are low, while the basal and peak LH and FSH responses to acute GnRH stimulation are not elevated. It is possible that transient minor elevations in oestradiol levels occur to which prepubertal breast tissue is exquisitely sensitive. Any exogenous supply of oestrogens, such as the application of oestrogen-containing creams for skin disorders, the accidental ingestion of an oral contraceptive tablet by the child or even consumption of foodstuffs (e.g. chicken) from animals pre-fed oestrogens, can all produce inappropriate breast development. A detailed history is clearly essential. Concentrations of sex hormone-binding globulin (SHBG) have been reported to be increased in patients with premature thelarche. The consequence of this change would be a decrease in free testosterone concentration with an increase in the ratio of free oestradiol to testosterone levels. This may explain the breast development, even though total oestradiol levels are not increased. Ultrasound studies are usually normal, although an occasional small ovarian cyst is found.

The natural history of premature thelarche is one of fluctuating breast development throughout childhood which does not interfere with the pattern and timing of the onset of true puberty. In a series of 48 cases studied long-term, menarche was normal, as was reproductive potential[9].

Early and late puberty

The causes of early and late puberty in girls are numerous and are documented in standard texts[10]. Early, true precocious puberty in girls is usually idiopathic in nature and is the result of premature activation of the GnRH pulse generator. It is hardly surprising that it remains idiopathic, while the precise trigger for the onset of normal puberty has still to be clarified. Endocrine investigation is seldom rewarding other than to confirm that gonadotrophin levels, particularly following acute stimulation with GnRH, are appropriate for a pubertal, rather than a prepubertal age. Plasma oestradiol levels are often, surprisingly, not too elevated. Ultrasonography offers a useful assessment, the dominant features being enlargement of the uterus (with an adult corpus to cervix size ratio) and ovaries. Large ovarian cysts are typically seen in the McCune-Albright syndrome, which is one cause of early puberty now thought to be due to autonomous ovarian hyperfunction. An absent or reduced gonadotrophin response to acute GnRH stimulation is the typical endocrine feature of this syndrome, whose other clinical signs include a characteristic pattern of skin pigmentation and cystic changes on bone radiographs due to polyostotic fibrous dysplasia.

Delayed puberty in females may present initially because of the absence of breast development, short stature, or even primary amenorrhoea. It is important to exclude Turner's syndrome before embarking on detailed endocrine and gynaecological investigations. Because plasma oestradiol levels are characteristically low in childhood, investigation of the cause is not helped by measurement of the steroid. Chronic disorders such as inflammatory bowel disease and the effects of anorexia nervosa are conditions which may sometimes be initially missed.

Information on the pulsatile nature of gonadotrophin secretion has been applied successfully to the treatment of both early and late puberty. Long-acting agonists of GnRH are widely available for therapeutic use in many clinical disorders, including gonadotrophin-dependent precocious puberty. The administration of GnRH in a continuous rather than in a pulsatile fashion leads to desensitization of the pituitary gonadotrophin receptors and a consequent decrease in circulating concentrations of LH and FSH. The response to treatment can be monitored by ultrasound, as illustrated in Figure 7, which shows the dramatic change in the appearance of large multiple cystic changes in the ovaries following treatment with GnRH analogue in a girl with precocious puberty. Conversely, giving GnRH in a pulsatile fashion is an effective form of treatment to induce puberty. This has been used in the treatment of simple constitutional delay in puberty and for the restoration of ovarian function in anorexia nervosa. However, the impracticality of using portable subcutaneous infusion pumps has meant that simple delayed puberty, at least, is generally managed either expectantly or with a short course of supplementary oestrogen treatment.

CONCLUSION

The ovary has generally been regarded as a quiescent endocrine organ before puberty. This is partly true, particularly when compared with its counterpart in the male sex. However, there is abundant evidence to indicate that the ovary does not lie dormant, either before or after birth. The most persuasive evidence for this comes from pelvic ultrasonography

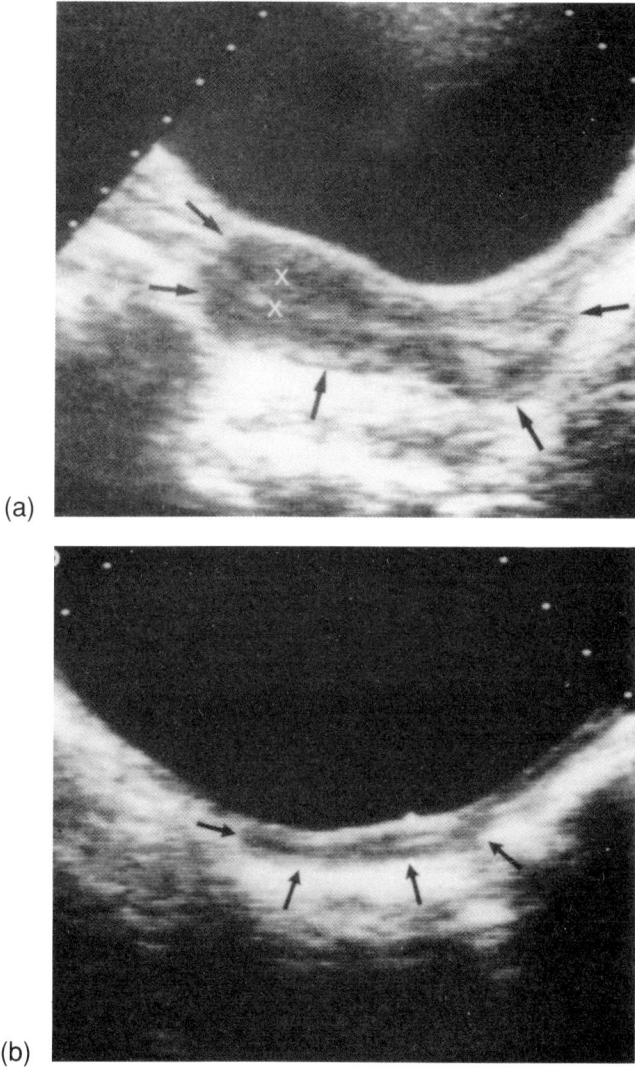

(a)

(b)

Figure 7 Sagittal midline scan of the uterus in a 6-year-old girl before (a) and after (b) treatment for precocious puberty with a GnRH analogue. Reproduced with permission from ref. 12

which, through its powers of sensitivity and detail, has provided valuable information for both diagnostic purposes and for monitoring treatment.

REFERENCES

1. Sinclair, A.H., Berta, P., Palmer, M.S., Hawkins, J.R., Griffiths, B., Smith, M., Foster, J., Frischauf, A-M., Lovell-Badge, R. and Goodfellow, P.N. (1990). A gene from the human sex-determining region encodes a protein with homology to a conserved DNA-binding motif. *Nature (London)*, **346**, 240–4

2. Tanner, J.M. (1962). *Growth at Adolescence*, 2nd edn. (Oxford: Blackwell Scientific Publications)

3. Eveleth, P.B. and Tanner, J.M. (1990). *Worldwide Variation in Human Growth*. (Cambridge: Cambridge University Press)

4. Voutilainen, R. and Miller, W.L. (1986). Developmental expression of genes for the steroidogenic enzymes P450 scc (20,22 desmolase), P450 c17 (17-hydroxylase/17,20 lyase) and P450 c21 (21-hydroxylase) in the human fetus. *J. Clin. Endocrinol. Metab.*, **63**, 1145–50

5. Grumbach, M.M., Roth, J.C., Kaplan, S.L. and Kelch, R.P. (1974). In Grumbach, M.M., Grave, G.D. and Mayer, F.E. (eds.). *The Control of the Onset of Puberty*, pp. 115–51. (New York: J. Wiley and Sons)

6. Wilson, D.W., Read, G.F., Hughes, I.A., Walker, R.F. and Griffiths, K. (1984). Hormone rhythms and breast cancer chronoepidemiology: salivary progesterone concentrations in pre- and post-menarchal girls and in normal premenopausal women. *Chronobiol. Int.*, **1**, 159–65

7. Polhemus, D.W. (1953). Ovarian maturation and cyst formation in children. *Pediatrics*, **11**, 588–94

8. Siegel, M.J. (1991). Pediatric gynecologic sonography. *Radiology*, **179**, 593–600

9. Van Winter, J.T., Noller, K.L., Zimmerman, D. and Melton, L.J. (1990). Natural history of premature thelarche in Olmsted County, Minnesota, 1940 to 1984. *J. Pediatr.*, **116**, 278–80

10. Stanhope, R. and Brook, C.G.D. (1989). Disorders of puberty. In Brook, C.G.D. (ed). *Clinical Paediatric Endocrinology*, pp. 189–212. (Oxford: Blackwell Scientific Publications)

11. Griffin, J.E. and Ojeda, S.R. (1988). *Textbook of Endocrine Physiology*, pp. 129–64. (Oxford: Oxford University Press)

12. Adams, J. (1989). The role of pelvic ultrasound in the management of paediatric endocrine disorders. In Brook, C.G.D. (ed.) *Clinical Paediatric Endocrinology*, pp. 675–91. (Oxford: Blackwell Scientific Publications)

5

Ovulatory dysfunction in endocrine disorders

P.G. Wardle and R. Fox

INTRODUCTION

Ovulatory dysfunction typically presents with menstrual disorders, infertility, miscarriage or hirsutism. Oligo- and amenorrhoea are the usual menstrual symptoms and may be the result of a variety of psychological, functional and anatomical defects. Although the patient's immediate concern is her presenting symptom(s), it is now clear that such women may also be at increased risk of several systemic complications in the longer term, including osteoporosis[1], diabetes mellitus[2], and cardiovascular disease[3].

Recent advances of our understanding of some of the pathophysiological mechanisms involved allows a simplified classification of ovulatory dysfunction from an ovarian standpoint. Three major categories can be defined: primary ovarian failure, secondary ovarian failure and polycystic ovarian disease. Table 1 shows the different relative contribution of these as causes of oligomenorrhoea or amenorrhoea.

PRIMARY OVARIAN FAILURE

General features

Primary ovarian failure is characterized by elevated serum gonadotrophin levels and oestrogen deficiency. It represents end-organ failure or

Table 1 Frequency (%) of endocrine disorders in anovulatory women with amenorrhoea or oligomenorrhoea

	Amenorrhoea	*Oligomenorrhoea*
Primary ovarian failure	10	2
Secondary ovarian failure		
hypothalamic	32	6
hyperprolactinaemia	22	2
Polycystic ovarian disease	33	88
Others	3	2

resistance of which there are several causes. Ovarian failure before the age of 40 has been estimated to affect 0.3–0.9% of women[4,5]. Aetiological factors include genetic abnormalities of the X-chromosome[6], environmental toxins (including ionizing radiation)[7], mumps infection[8], metabolic diseases (e.g. galactosaemia[9]) and autoimmune anomalies[10]. However, in the majority of women who present with secondary amenorrhoea no underlying cause is found.

Although some early studies showed the invariable absence of viable follicles on ovarian biopsy in all women with hypergonadotrophic amenorrhoea[11], there were other workers whose reports indicated that some cases of primary ovarian failure were reversible[12,13]. The terms 'insensitive' or 'gonadotrophin-resistant' ovarian syndrome have been used to describe such women who frequently have normal numbers of primordial follicles on ovarian biopsy. Pregnancies have been reported spontaneously (infrequently)[13,14], following high-dose gonadotrophin therapy[15], glucocorticoids[16,17] and oestrogens, either alone[18,19] or followed by human menopausal gonadotrophin[20]. Some pregnancies have occurred in women when no follicles were seen at prior ovarian biopsy[21].

Pathophysiology

Biopsy findings in women with gonadotrophin-resistant ovary syndrome demonstrate an arrest following early development of primordial follicles.

Table 2 Results of oestrogen–human menopausal gonadotrophin (hMG) and gonadotrophin releasing hormone (GnRH) analogue–hMG treatment in amenorrhoeic women with hypergonadotrophic hypogonadism (adapted from ref. 20)

	Oestrogen–hMG	*GnRH analogue–hMG*
Patients	91	9
Cycles	311	43
Patients ovulating	34	3
Ovulations	61 (20%)	7 (16%)
Pregnancies	19	0
Miscarriages	10	0
Stillbirths	1	0
Viable births	8	0

It is recognized that development to this stage can occur with minimal or no gonadotrophin stimulation[22]. Normally, the granulosa cells produce oestradiol which stimulates development of follicle stimulating hormone (FSH) receptors[23]. FSH and oestradiol act together to induce luteinizing hormone (LH) receptors and antral formation occurs with further growth of the follicle. Failure of these mechanisms in women with gonadotrophin-resistant ovary syndrome is not due to a biologically inactive FSH molecule[24]; the molecular weight and FSH sialic acid content are similar to that of normal postmenopausal women[25], antibodies to LH and FSH have not been demonstrated[25,26] and intrafollicular antigonadotrophic action of prolactin is unlikely, as serum prolactin levels are not raised.

Various possible causes of follicular unresponsiveness have been postulated, including defective gonadotrophins, defective or absent gonadotrophin receptors, receptor blocking factors or impaired postreceptor cellular responses[10]. The recent studies by Check and colleagues[20] used either oestrogens or a gonadotrophin-releasing hormone analogue (leuprolide acetate), followed by human menopausal gonadotrophin (hMG) stimulation to induce ovulation and their results are summarized in Table 2. Appropriate follicular development occurred and ovulation was induced with equivalent frequency (16–20%) in both groups.

Interestingly, pregnancies were only achieved in the group treated with oestrogen priming followed by hMG. The successful results suggest a number of mechanisms whereby oestrogen or gonadotrophin releasing hormone (GnRH) analogues might restore ovarian responsiveness. Ovarian gonadotrophin receptors may be down-regulated and oestrogen may restore responsiveness either by a direct effect or indirectly by negative feedback to the hypothalamus and pituitary, lowering elevated gonadotrophin levels. It is also possible that oestrogens might enhance FSH binding to receptors, as has been suggested in rat animal models[27,28] and render subsequent additional hMG stimulation more effective than when GnRH analogues are used to suppress endogenous gonadotrophins. The effect of GnRH analogue may be mediated by a priming effect from its initial agonist action on ovarian gonadotrophin receptors.

SECONDARY OVARIAN FAILURE

General features

Secondary ovarian failure, resulting from hypothalamic or pituitary failure or disorder (including hyperprolactinaemia), is characterized by non-elevated LH and FSH levels, and oestrogen deficiency. LH (and presumably GnRH) pulsatility revert to pubertal or pre-pubertal patterns which may be reflected by, respectively, a multifollicular or a totally inactive appearance with ovarian ultrasonography. Hyperprolactinaemia and hypothalamic functional disorders (including weight loss-, exercise- and stress-related amenorrhoea) are responsible for the great majority of cases.

Hyperprolactinaemia

Prolactin is unique amongst the anterior pituitary hormones in that its secretion is predominantly under tonic inhibitory control by a hypothalamic factor – dopamine. If this physiological mechanism is disrupted due to section of the pituitary stalk, prolactin is secreted in excess. Other pathological causes of hyperprolactinaemia include prolactinergic drug therapy such as phenothiazines and metoclopramide, primary hypothyroidism, and pituitary prolactinomas.

As in women with weight loss, over 90% of women with hyperprolactinaemic amenorrhoea have oestrogen deficiency[29,30]. Serum concentrations of LH and FSH are generally within the normal range for the early follicular phase[30]. These data led some workers to suppose that prolactin had a direct inhibitory effect on the ovary *in vivo*, blocking the action of gonadotrophins[31]. Using granulosa cells in culture as a model to study the effect of prolactin on LH-dependent steroidogenesis, McNatty and colleagues[32] appeared to show that high concentrations of prolactin inhibited progesterone synthesis. The relevance of this *in vitro* observation to the development of anovulation in hyperprolactinaemia is unclear, however. Prolactin concentrations have been found to be significantly lower in conception than non-conception cycles in ovulatory infertile women with hyperprolactinaemia[33] but a controlled trial has failed to demonstrate benefit from prolactin suppression with bromocriptine[34]. The hypothesis that hyperprolactinaemia *per se* directly inhibits ovarian function fails to recognize that the levels of FSH and LH are inappropriately low, given a state of oestrogen deficiency, and that the ovarian response to exogenous gonadotrophin therapy in hyperprolactinaemia is normal[35].

It has been postulated that prolactin has a central effect on gonadotrophin secretion. Impaired secretion of gonadotrophins in women with large tumours can be easily explained by compression of the pituitary or its stalk. Indeed, this may occasionally result in panhypopituitarism. However, in the majority of women who have microadenomas or idiopathic hyperprolactinaemia, impairment of FSH and LH is the result of a *functional* defect. Boyar and co-workers[36] showed that LH pulse frequency and amplitude are reduced in hyperprolactinaemia, and Glass and colleagues[37] have demonstrated a lack of positive feedback to exogenous oestrogen. Interestingly, Polson and colleagues[38] successfully induced ovulation with pulsatile GnRH in five hyperprolactinaemic women who failed to respond to, or were intolerant of, the dopamine agonist bromocriptine, indicating that the principal mechanism of anovulation in hyperprolactinaemia is the disordered release of GnRH. Curiously, in contrast to women with weight loss-related amenorrhoea, women with hyperprolactinaemic amenorrhoea have a normal or even exaggerated response to GnRH[30,37].

Weight loss-related amenorrhoea

Weight loss of 10–15% in normally proportioned, eumenorrhoeic women results in secondary amenorrhoea. The menstrual disturbance does not seem to be related to the absolute weight or proportion of fat, but rather to the amount of weight lost, expressed as a percentage of the pre-morbid weight. Some authors have speculated that weight loss-related amenorrhoea is a biological adaption, avoiding the birth of young during times of great hardship. The altered body composition has wide-ranging effects throughout the endocrine system, including abnormalities of thyroid function, growth hormone secretion, and the hypothalamic–pituitary–adrenal axis (reviewed by Van der Spuy[39]).

Basal serum FSH and LH levels are low[40]. LH concentrations are generally more affected than those of FSH, and LH pulsatility is lost. Serum oestradiol concentrations are low normal and at ultrasound scan the ovaries are generally small and inactive or multifollicular in appearance. The uterine volume is markedly reduced, the endometrium is atrophic and the menstrual response to progestogen challenge is negative or scanty[29,41], all reflecting an oestrogen-deficient state.

The pituitary gonadotrophin response to an intravenous bolus of GnRH is blunted, but evidence that pulsatile GnRH successfully induces ovulation in the majority of underweight women indicates that the pituitary and ovary are functionally intact[42]. The composite picture, therefore, is one of hypogonadotrophic hypogonadism with ovarian inactivity secondary to an acquired dysfunction of the hypothalamic GnRH pulse generator.

In addition to the effects of hypogonadism on oestrogen state, there is evidence that the decrease in body fat limits peripheral aromatization of circulating androgens to oestrogen. Moreover, with extreme emaciation, oestradiol undergoes metabolism by 2-hydroxylation (instead of the 16-hydroxylation pathway) with formation of the catechol oestrogen, 2-hydroxy-oestrone, which has no intrinsic oestrogen activity[43] and therefore amplifies negative rather than positive feedback.

Active oestrogens inhibit the secretion of endogenous opioids in the arcuate nucleus of the hypothalamus and oestrogen deficiency withdraws this inhibition. Increased inactive oestrogen metabolites (such as 2-hydroxy-oestrone) can also interfere with this inhibition by competitive blockade. Endogenous opioids such as β-endorphin are known to exert a tonic inhibition of pulsatile GnRH as shown from *in vitro* studies of human hypothalamic responses using naloxone which blocks opiate receptors and thereby permits GnRH release[44]. Oestrogen deficiency or increased levels of inactive oestrogen metabolites may thus allow increased secretion of endogenous opioids which inhibit GnRH pulsatility. In addition, corticotrophin-releasing factor (CRF) stimulates endogenous opioid secretion. Amenorrhoea associated with psychological stress, where CRF and day-time cortisol levels are raised, may also be mediated by opioid effects in this manner.

Although there is concordance of changes in body weight and neuroendocrine function, there are disparities in a proportion of patients, particularly those with anorexia nervosa. A significant proportion of anorexic women became amenorrhoeic before losing a substantial amount of weight, and many do not resume menstruation when their normal weight is regained. These observations suggest contributory psychogenic factors, which are well recognized in anorexia nervosa. These may be mediated via corticosteroid and other stress hormone changes which may act synergistically with weight loss. This creates some uncertainty about the primary role of body weight with respect to ovarian function.

Exercise-related amenorrhoea

Strenuous physical exercise has become increasingly popular during the past decade and with it an increased awareness of the potential adverse effects on ovarian function. The prevalence of amenorrhoea is much greater in distance runners than in swimmers[45] and this may be attributable to the relative low percentage of body fat in runners (15%) compared with swimmers (20%). Although amenorrhoeic runners exhibit similar alterations in basal FSH and LH levels and LH pulsatility to those seen in women with weight loss-related amenorrhoea, the mechanisms involved may be different. Compared with eumenorrhoeic runners, amenorrhoeic runners have an exaggerated response to a GnRH bolus[46], unlike the blunted response in women with weight loss-related amenorrhoea. This pattern is similar to that reported in psychogenic or stress-related amenorrhoeic women[47,48].

Analysis of the mechanisms responsible for exercise-related amenorrhoea is complicated by the potential synergistic effects of weight loss and psychological factors in amenorrhoeic runners. Sanborn and colleagues[45], in a questionnaire study of 237 runners, showed that the risk of amenorrhoea was closely related to the women's weight (Figure 1). Gadpaille and colleagues[49] demonstrated an increased frequency of eating disorders and of a personal or family history of major affective disorder in amenorrhoeic compared to normally menstruating runners. The results of their study are summarized in Table 3. As previously mentioned, stress is associated with increased production of CRF which may inhibit pulsatile GnRH release via its effects on endogenous opioids. This may be important in the light of Gadpaille's observations of psychogenic disorder[49].

In addition, it is recognized that exercise itself leads to transient increases in serum prolactin[50] and cortisol concentrations[51]. Studies comparing the response in amenorrhoeic and normally menstruating runners are contradictory, but superimposed stress effects may act synergistically and produce amenorrhoea, or those who develop this may have an enhanced sensitivity to these exercise-induced changes. However, acute exercise in runners who are already amenorrhoeic runners does not induce a prolactin response, which suggests this is unlikely to be the mediator of GnRH and gonadotrophin suppression.

Physical stress, rather than weight change, is undoubtedly a principal factor, irrespective of how this is mediated, because LH pulsatility and menstruation return rapidly, even without any weight change, when training ceases abruptly.

POLYCYSTIC OVARIAN DISEASE

General features

The classic report of polycystic ovarian disease (PCOD) described women with obesity, amenorrhoea, and severe hyperandrogenaemia. These criteria define but a small proportion of women with polycystic ovaries (PCOs) however. It is now clear that PCOs are much more common than previously recognized, and can be found in a significant proportion of apparently normal women[41,52,53].

The characteristic endocrine profile of anovulatory women with PCOs consists of a non-elevated serum concentration of FSH, elevated serum concentrations of LH and testosterone, and a subnormal serum concen-

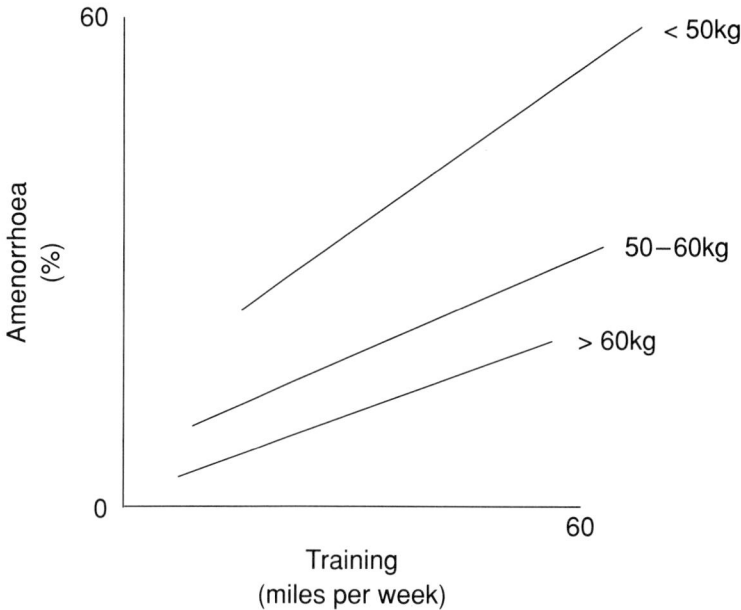

Figure 1 The prevalence of amenorrhoea in runners correlated inversely with weight and directly with training intensity (adapted from ref. 45, with permission)

tration of sex hormone-binding globulin. The consensus is that the excess androgen production is entirely ovarian in origin in at least 80% of cases. In the remaining cases there is an additional component from the adrenal glands. Although more than 90% of anovulatory women with PCOs will have an elevated free androgen index[41,54], only in fewer than half the cases is this expressed as clinical hyperandrogenism.

Despite the clear evidence of hyperandrogenaemia in the vast majority of cases, over 90% of anovulatory women with PCOs are oestrogenized[41]. Data from selective catheterization studies have shown that the ovary is the predominant source of oestradiol[55]. Although individual follicles in the polycystic ovary are relatively oestrogen-deficient[56], in combination the numerous follicles secrete substantial quantities of oestradiol.

Anovulatory women with PCOs and oestrogen deficiency fall into two groups: severely androgenized women, and non-hirsute women with weight

Table 3 Eating, major affective disorders and family history in 13 amenorrhoeic and 19 normally menstruating runners (adapted from ref. 49)

	Amenorrhoeic		Normally menstruating	
	n	%	*n*	%
Eating disorders	8	62	0	0★★
Major affective disorder	3	23	0	0
1st/2nd-degree relative with affective disorder	10	77	1	5★★

★★ p < 0.001

loss in whom the polycystic ovary is an incidental finding which is irrelevant to the presenting disorder[41]. Polycystic ovaries have also been found in women with hyperprolactinaemic amenorrhoea[41,57,58]. In these cases it appears that the hypothalamic disorder overrides pre-existing PCOD.

The combined disturbance of pituitary and ovarian function gave rise to the hypothesis of a self-perpetuating vicious circle[59], the ovarian dysfunction being dependent on LH hypersecretion, and vice versa. Androgen production is manifestly dependent upon LH[60], and ovarian surgery is able to modulate gonadotrophin release[53]. Sustained suppression of LH with GnRH agonists not only has little effect on ovarian morphology, but also, following withdrawal, spontaneous ovulation is rare. These observations suggest that the polycystic ovary is not simply a spontaneous aberration of hypothalamic–pituitary–ovarian control, but that a defect exists which hinders the establishment of normal homeostatic mechanisms. In addition, the finding of PCOs in ovulatory women[52] indicates that a polycystic morphology of the ovary is not simply a non-specific end-result of anovulation.

Insulin resistance and adrenal hyperandrogenaemia (AH) are well-recognized features of PCOD, and have been proposed as aetiological factors. The mechanisms for the development of insulin resistance and AH are poorly understood and, more importantly, it is not known if they represent two discrete endocrinopathies or different facets of a single disorder.

Hyperinsulinaemia

Insulin resistance and hyperinsulinaemia are now well-recognized features of PCOD in both obese[61,62] and non-obese patients[63,64]. There has been great controversy over the nature of this association and whether it is the result or the cause of hyperandrogenism. Evidence that hyperandrogenism causes increased insulin resistance comes from studies by Cole and Kitabchi[65] who reported a case of amelioration of insulin resistance by oestrogen therapy, and Woodard and colleagues[66], who described the development of insulin resistance in young boys given synthetic androgen for aplastic anaemia. In addition, Shoupe and Lobo[67] gave the androgen antagonist, spironolactone, to women with PCOs and were able to reduce fasting insulin levels as well as testosterone concentrations. However, the mechanism of insulin resistance in ovarian hyperandrogenism has been recognized to vary from case to case[68], and the observation that suppression of ovarian androgen production using a long-acting LHRH agonist has no effect on glucose-stimulated insulin concentrations[69] appears to indicate that hyperinsulinaemia is more likely to be the cause of hyperandrogenism rather than a result of it.

Hyperinsulinaemia may affect ovarian function directly by augmenting LH-dependent androgen production[70,71], and indirectly by amplifying GnRH-dependent LH production by the pituitary[72]. In addition, it may also act by suppressing hepatic production of sex hormone-binding globulin[73].

Adrenal hyperandrogenaemia

The observation that women with congenital adrenal hyperplasia have an increased prevalence of PCOs at ultrasound scan[74] seems to point clearly to PCOD arising in response to a discrete abnormality of adrenal androgenesis. Evidence from studies of primate cell cultures[75] of an atretic action of testosterone on granulosa cells provided a simple paradigm for the induction of ovarian dysfunction by adrenal hyperandrogenaemia. However, testosterone implants do not appear to influence ovarian function in women with normal menstrual cycles[76]. It now seems that in the majority of cases, AH is but another feature of insulin resistance[77].

Potentially therefore, insulin has wide-ranging effects upon the endocrine system, with the ability to disrupt pituitary, ovarian, and adrenal function. It seems likely that any condition which directly or indirectly increases insulin resistance will adversely affect folliculogenesis. Such conditions include untreated hyperthyroidism, Cushing's syndrome, acromegaly, and treated insulin-dependent diabetes mellitus.

CONCLUSIONS

Recent advances in our understanding of the ovarian cycle and mechanisms of ovulatory failure allow a simplified classification of anovulatory disorders according to oestrogen state and FSH concentrations:

(1) Primary ovarian failure, characterized by elevated FSH levels and oestrogen deficiency, represents end-organ resistance of which there are several causes including anti-ovarian antibodies, radiotherapy, and karyotype abnormalities.

(2) Secondary ovarian failure, resulting from hypothalamic or pituitary failure or disorder including hyperprolactinaemia, is typified by non-elevated FSH levels and oestrogen deficiency. At ultrasonography the ovaries appear totally inactive or have a multifollicular appearance. Occasionally the acquired abnormality will be overriding pre-existing polycystic ovaries.

(3) Polycystic ovarian disease, the result of a number of distinct endocrinopathies including hyperinsulinaemia and adrenal hyperandrogenaemia, is typified by non-elevated FSH and oestro-genization of the endometrium.

REFERENCES

1. Davies, M.C., Hall, M.L. and Jacobs, H.S. (1990). Bone mineral loss in young women with amenorrhoea. *Br. Med. J.*, **301**, 790–3
2. Dahlgren, E., Johansson, S., Lindstedt, G., Knutsson, F., Olden, A., Janson, P., Mattson, L-A., Crona, N. and Lundberg, P-A. (1992). Women with polycystic ovary syndrome wedge resected in 1956 to 1965: a long-term follow-up focusing on natural history and circulating hormones. *Fertil. Steril.*, **57**, 505–13
3. Wild, R.A., Grubb, B., Hartz, A., Van Nort, J.J., Bachman, W. and

Bartholomew, M. (1990). Clinical signs of androgen excess as risk factors for coronary artery disease. *Fertil. Steril.*, **45**, 255–9

4. Aiman, J. and Smentek, C. (1985). Premature ovarian failure. *Obstet. Gynecol.*, **66**, 9–14

5. Coulam, C.B., Adamson, S.C. and Annegers, J.F. (1986). Incidence of premature ovarian failure. *Obstet. Gynecol.*, **67**, 604–6

6. Simpson, J.L. (1987). Phenotypic–karyotypic correlations of gonadal determinants: Current status and relationship to molecular studies. In Vogel, F. and Sperling, K. (eds.) *Proceedings of the International Congress of Human Genetics*, pp. 224–32, (Heidelberg: Springer-Verlag)

7. Verp, M.S. (1983). Environmental causes of ovarian failure. *Semin. Reprod. Endocrinol.*, **1**, 101–11

8. Morrison, J.C., Givens, J.R., Wiser, W.L. and Fish, S.A. (1975). Mumps oophritis: A cause of premature menopause. *Fertil. Steril.*, **26**, 655–9

9. Chen, Y-T., Mattison, D.R., Feigenbaum, L., Fukui, H. and Schulman, J.D. (1981). Reduction in oocyte number following prenatal exposure to a diet high in galactose. *Science*, **214**, 1145–7

10. La Barbera, A.R., Miller, M.M., Ober, C. and Rebar, R.W. (1988). Autoimmune aetiology in premature ovarian failure. *Am. J. Reprod. Immunol. Microbiol.*, **16**, 115–22

11. Goldenberg, R.L., Grodin, J.M., Rodbard, D. and Ross, G.T. (1973). Gonadotropins in women with amenorrhoea. *Am. J. Obstet. Gynecol.*, **116**, 1003–12

12. O'Herlihy, C., Pepperell, R.J. and Evans, J.H. (1980). The significance of FSH elevation in young women with disorders of ovulation. *Br. Med. J.*, **281**, 1447–50

13. Rebar, R.W., Erickson, G.F. and Yen, S.S.C. (1982). "Idiopathic premature ovarian failure": clinical and endocrine characteristics. *Fertil. Steril.*, **37**, 35–41

14. Schreiber, J.R., Davajan, V. and Kletzky, O.A. (1978). A case of intermittent ovarian failure. *Am. J. Obstet. Gynecol.*, **132**, 698–9

15. Johnson, T.R. and Peterson, E.R. (1979). Gonadotropin-induced pregnancy following "premature ovarian failure". *Fertil. Steril.*, **31**, 351–2

16. Meldrum, D.R., Frumar, A.M., Shamonki, I.M., Benirschke, K. and Judd, H.L. (1980). Ovarian and adrenal steroidogenesis in a virilised patient with gonadotropin-resistant ovaries and hilus cell hyperplasia. *Obstet. Gynecol.*, **56**, 216–21

17. Coulam, C.M., Kempers, R.D. and Randall, R.V. (1981). Premature ovarian failure: evidence for the autoimmune mechanism. *Fertil. Steril.*, **36**, 238–40

18. Starup, J., Philip, J. and Sele, V. (1978). Oestrogen treatment and

subsequent pregnancy in two patients with severe hypergonadotrophic ovarian failure. *Acta Endocrinol.*, **89**, 149–57

19. Szlachter, B.N., Nachtigall, L.E., Epstein, J., Young, B.K. and Weiss, G. (1979). Premature menopause: a reversible entity. *Obstet. Gynecol.*, **54**, 396–8

20. Check, J.H., Nowroozi, K., Chase, J.S., Nazari, A., Shapse, D. and Vase, M. (1990). Ovulation induction and pregnancies in 100 consecutive women with hypergonadotropic amenorrhoea. *Fertil. Steril.*, **53**, 811–16

21. Rebar, R.W. and Connolly, H.V. (1990). Clinical features of young women with hypergonadotropic amenorrhoea. *Fertil. Steril.*, **53**, 804–10

22. Goldenberg, R.L., Powell, R.D., Rosen, S., Marshall, J. and Ross, G.T. (1976). Ovarian morphology in women with anosmia and hypogonadotropic hypogonadism. *Am. J. Obstet. Gynecol.*, **126**, 91–4

23. Richards, J.S. (1980). Maturation of ovarian follicles: actions and interactions of pituitary and ovarian hormones on follicular differentiation. *Physiol. Rev.*, **60**, 51–89

24. Jones, G.S. and De Moraes-Ruehsen, M. (1969). A new syndrome of amenorrhoea in association with hypergonadotropism and apparently normal ovarian follicular apparatus. *Am. J. Obstet. Gynecol.*, **104**, 597–600

25. Koninckx, P.R. and Brosens, I.A. (1977). The "gonadotropin-resistant ovary" syndrome as a cause of secondary amenorrhoea and infertility. *Fertil. Steril.*, **28**, 926–31

26. Starup, J. and Pedersen, J. (1978). Hormonal and ultrastructural observations in a case of resistant-ovary syndrome. *Acta Endocrinol.*, **89**, 744–52

27. Louvet, J.P. and Vaitkuaitis, J.L.(1976). Induction of follicle-stimulating hormone (FSH) receptors in rat ovaries by oestrogen priming. *Endocrinology*, **99**, 758–64

28. Richards, J.S., Ireland, J.J. and Rao, M.C. (1976). Ovarian follicular development in the rat: hormone receptor regulation by estradiol, follicle-stimulating hormone and luteinizing hormone. *Endocrinology*, **99**, 1562–70

29. Hull, M.G.R., Knuth, U.A., Murray, M.A.F. and Jacobs, H.S. (1979). The practical value of the progestogen challenge test, serum oestradiol estimation or clinical examination in assessment of the oestrogen state and response to clomiphene in amenorrhoea. *Br. J. Obstet. Gynaecol.*, **86**, 799–805

30. Jacobs, H.S., Franks, S., Murray, M.A.F., Hull, M.G.R., Steele, S.J. and Nabarro, J.D.N. (1976). Clinical and endocrine features of hyperprolactinaemic amenorrhoea. *Clin. Endocrinol.*, **5**, 439–54

31. Thorner, M.O. (1977). Prolactin. *Clin. Endocrinol. Metab.*, **6**, 201–22

32. McNatty, K.P., Sawers, R.S. and McNeilly, A.S. (1974). A possible role for

prolactin in control of steroid secretion by the human Graafian follicle. *Nature (London)*, **250**, 653–5

33. Lenton, E.A., Brook, L.M., Sobowale, O. and Cooke, I.D. (1979). Prolactin concentrations in normal menstrual cycles and conception cycles. *Clin. Endocrinol.*, **10**, 383–91

34. Wright, C.W., Steele, S.J. and Jacobs, H.S.(1979). Value of bromocriptine in unexplained primary infertility: a double-blind controlled trial. *Br. Med. J.*, **1**, 1037–9

35. Fraser, I.S., Markham, R. and Shearman, R.P. (1978). Plasma prolactin levels and ovarian responsiveness to exogenous gonadotrophins. *Obstet. Gynecol.*, **51**, 548–51

36. Boyar, R.M., Katz, J., Finkelstein, J.W., Kapen, S., Weiner, H., Weltzman, E.D. and Hellman, L. (1974). Anorexia nervosa: Immaturity of the 24-hour luteinizing hormone secretory pattern. *N. Engl. J. Med.*, **291**, 861–5

37. Glass, M.R., Shaw, R.W., Williams, J.W., Butt, W.R., Logan-Edwards, R. and London, D.R. (1976). The control of gonadotrophin release in women with hyperprolactinaemic amenorrhoea: effect of oestrogen and progesterone on the LH and FSH response to LHRH. *Clin. Endocrinol.*, **5**, 521–30

38. Polson, D.W., Sagle, M., Mason, H.D., Adams, J., Jacobs, H.S. and Franks, S. (1986). Ovulation and normal luteal function during LHRH treatment of women with hyperprolactinaemic amenorrhoea. *Clin. Endocrinol.*, **24**, 531–7

39. Van der Spuy, Z.M. (1985). Nutrition and reproduction. In Jacobs, H.S. (ed.) Reproductive Endocrinology. *Clinics in Obstetrics and Gynaecology*, vol. 12, pp. 579–604. (London: W.B. Saunders)

40. Knuth, U.A., Hull, M.G.R. and Jacobs, H.S. (1977). Amenorrhoea and loss of weight. *Br. J. Obstet. Gynaecol.*, **84**, 801–7

41. Fox, R., Corrigan, E., Thomas, R. and Hull, M.G.R. (1991). Oestrogen and androgen states in oligo-amenorrhoeic women with polycystic ovaries. *Br. J. Obstet. Gynaecol.*, **98**, 294–304

42. Nillius, S.J. and Wide, L. (1979). Effects of prolonged luteinizing hormone releasing-hormone therapy on follicle maturation, ovulation and corpus luteum function in amenorrheic women with anorexia nervosa. *Ups. J. Med. Sci.*, **84**, 21–35

43. Fishman, J. and Bradlow, H.L. (1976). Effect of malnutrition on the metabolism of sex hormones in man. *Clin. Pharmacol. Therap.*, **22**, 721–8

44. Rasmussen, D.D., James, H., Lui, J.H. and Yen, S.S.C. (1983). Endogenous opioid regulation of GnRH release from the human medio-basal hypothalamus (MBH) *in vitro*. *Fertil. Steril.*, **40** (Abstr.), 418

45. Sanborn, C.F., Martin, B.J. and Wagner, W.W. (1982). Is athletic

amenorrhea specific to runners? *Am. J. Obstet. Gynecol.*, **143**, 859–61

46. Yahiro, J., Glass, A.R., Fears, W.B., Ferguson, E.W. and Vigersky, R.A. (1987). Exaggerated gonadotropin response to luteinizing hormone-releasing hormone in amenorrhoeic runners. *Am. J. Obstet. Gynecol.*, **156**, 586–91

47. Yaginuma, T. (1979). Progress and therapy of stress amenorrhoea. *Fertil. Steril.*, **32**, 36–9

48. Anapliotou, M., Panitsa-Faflia, C., Pitoulis, S., Lipraki, M. and Batrinos, M.L. (1982). Study of the pituitary gonadotropin pulses and the positive feedback mechanism of LH release in psychogenic and postpartum amenorrhea. *Acta Endocrinol.*, **248** (suppl.), 5–6

49. Gadpaille, W.J., Sanborn, C.F. and Wagner, W.W. (1987). Athletic amenorrhea, major affective disorders and eating disorders. *Am. J. Psych.*, **144**, 939–42

50. Chang, F.E., Dodds, W.G., Sullivan, M., Kim, M.H. and Markey, W.B. (1986). The acute effects of exercise on prolactin and growth hormone secretion: comparison between sedentary women and women runners with normal and abnormal menstrual cycles. *J. Clin. Endocrinol. Metab.*, **62**, 551–6

51. Villanueva, A.L., Schlosser, C., Hopper, B., Lui, J.H., Hoffman, D.I. and Rebar, R.W. (1986). Increased cortisol production in women runners. *J. Clin. Endocrinol. Metab.*, **63**, 133–6

52. Polson, D.W., Wadsworth, J., Adams, J. and Franks, S. (1988). Polycystic ovaries – a common finding in normal women. *Lancet*, **1**, 870–2

53. Abdel Gadir, A., Khatim, M.S., Mowafi, R.S., Alnaser, H.M.I., Alzaid, H.G.N. and Shaw, R.W. (1990). Hormonal changes in patients with polycystic ovarian disease after ovarian electrocautery or pituitary desensitization. *Clin. Endocrinol.*, **32**, 749–54

54. Eden, J.A., Place, J., Carter, G.D., Jones, J., Alaghband-Zadeh, J. and Pawson, M.E. (1989). The diagnosis of polycystic ovaries in subfertile women. *Br. J. Obstet. Gynaecol.*, **96**, 809–15

55. Wajchenberg, B.L., Achando, S.S., Mathor, M.M., Czeresnia, C.E., Neto, D.G. and Kirschner, M.A. (1988). The source(s) of oestrogen production in hirsute women with polycystic ovarian disease as determined by simultaneous adrenal and ovarian venous catheterization. *Fertil. Steril.*, **49**, 56–61

56. Erickson, G.R., Hsueh, A.J.W., Quigley, M.E., Rebar, R.W. and Yen, S.S.C. (1979). Functional studies of aromatase activity in human granulosa cells from normal and polycystic ovaries. *J. Clin. Endocrinol. Metab.*, **49**, 514–19

57. Conway, G.S., Honour, J.W. and Jacobs, H.S. (1989). Heterogeneity of the

polycystic ovary syndrome: clinical, endocrine and ultrasound features in 556 patients. *Clin. Endocrinol.*, **30**, 459–70

58. Abdel Gadir, A., Khatim, M.S., Mowafi, R.S., Alnaser, H.M.I., Muharib, N.S. and Shaw, R.W. (1992). Implications of ultrasonically diagnosed polycystic ovaries. 1. Correlations with basal hormonal profiles. *Hum. Reprod.*, **4**, 453–7

59. Yen, S.S.C. (1980). The polycystic ovary syndrome. *Clin. Endocrinol.*, **12**, 177–208

60. Chang, R.J., Laufer, L.R., Meldrum, D.R., DeFazio, J., Lu, J.K.H., Vale, W.W., Rivier, J.E. and Judd, H.L. (1983). Steroid secretion in polycystic ovarian disease after ovarian suppression by a long-acting gonadotropin-releasing hormone agonist. *J. Clin. Endocrinol. Metab.*, **56**, 897–903

61. Burghen, G.A., Givens, J.R. and Kitabchi, A.E. (1980). Correlation of hyperandrogenism with hyperinsulinism in polycystic ovarian disease. *J. Clin. Endocrinol. Metab.*, **50**, 113–16

62. Pasquali, R., Venturoli, S., Paradis, R., Capelli, M., Parenti, M. and Melchionda, N. (1982). Insulin and C-peptide levels in obese patients with polycystic ovaries. *Horm. Metab. Res.*, **14**, 284–7

63. Chang, R.J., Nakamura, R.M., Judd, H.L. and Kaplan, S.A. (1983). Insulin resistance in non-obese patients with polycystic ovarian disease. *J. Clin. Endocrinol. Metab.*, **57**, 356–9

64. Jialal, I., Naiker, P., Reddi, K., Moodley, J. and Joubert, S.M.(1987). Evidence for insulin resistance in non-obese patients with polcystic ovarian disease. *J. Clin. Endocrinol. Metab.*, **64**, 1066–9

65. Cole, C. and Kitabchi, A.E. (1978). Remission of insulin resistance with Orthonovum in a patient with polycystic ovarian disease and acanthosis nigricans. *Clin. Res.*, **26**, (Abstr.) 412A

66. Woodard, T.L., Burghen, G.A., Kitabchi, A.E. and Williams, J.A. (1981). Glucose intolerance and insulin resistance in aplastic anemia treated with oxymetholone. *J. Clin. Endocrinol. Metab.*, **53**, 905–8

67. Shoupe, D. and Lobo, R.A. (1984). The influence of androgens on insulin resistance. *Fertil. Steril.*, **41**, 385–8

68. Bar, R.S., Muggeo, M., Roth, J., Kahn, C.R., Havrankova, J. and Imperato-McGinley, J. (1978). Insulin resistance, acanthosis nigricans and normal insulin receptors in a young woman: evidence for a postreceptor defect. *J. Clin. Endocrinol. Metab.*, **47**, 620–5

69. Geffner, M.E., Kaplan, S.A., Bersch, N., Golde, D.W., Landaw, E.M. and Chang, R.J. (1986). Persistence of insulin resistance in polycystic ovarian disease after inhibition of ovarian steroid secretion. *Fertil. Steril.*, **45**, 327–33

70. Adashi, E.Y., Resnick, C.E., Brodie, A.M.H., Svoboda, M.E. and Van Wyk, J.J. (1985). Somatomedin-C-mediated potentiation of follicle-

stimulating hormone-induced aromatase activity of cultured rat granulosa cells. *Endocrinology*, **117**, 2313–20

71. Barbieri, R.L., Makris, A. and Ryan, K.J. (1984). Insulin stimulates androgen accumulation in incubations of human ovarian stroma and theca. *Obstet. Gynecol.*, **64** (suppl.), 73–80S

72. Adashi, E.Y., Hsueh, A.J.W. and Yen, S.S.C. (1987). Insulin enhancement of luteinizing hormone and follicle-stimulating hormone release by cultured pituitary cells. *Endocrinology*, **108**, 1441–5

73. Kiddy, D.S., Hamilton-Fairley, D., Seppälä, M., Koistinen, R., James, V.H.T., Reed, M.J. and Franks, S. (1989). Diet-induced changes in sex hormone binding globulin and free testosterone in women with normal or polycystic ovaries: correlation with serum insulin and insulin-like growth factor-1. *Clin. Endocrinol.*, **31**, 757–63

74. Hague, W.M., Adams, J., Rodda, C., Brook, C.G.D., De Bruyn, R., Grant, D.B. and Jacobs, H.S. (1990). The prevalence of polycystic ovaries in patients with congenital adrenal hyperplasia and their close relatives. *Clin. Endocrinol.*, **33**, 501–10

75. Harlow, C.R., Hillier, S.G. and Hodges, J.K. (1986). Androgen modulation of follicle-stimulating hormone-induced granulosa cell steroidogenesis in the primate ovary. *Endocrinology*, **119**, 1403–5

76. Dewis, P., Newman, M., Ratcliffe, W.A. and Anderson, D.C. (1986). Does testosterone affect the normal menstrual cycle? *Clin. Endocrinol.*, **24**, 515–21

77. Lanzone, A., Fulghesu, A.M., Guido, M., Fortini, A., Caruso, A. and Mancuso, S. (1992). Differential androgen response to adrenocorticotropic hormone stimulation in polycystic ovarian syndrome: relationship with insulin secretion. *Fertil. Steril.*, **58**, 296–301

6

Purified gonadotrophin preparations for induction of ovulation

S. Franks, D.W. Polson, M. Sagle, D. Hamilton-Fairley, D.S. Kiddy and H.D. Mason

INTRODUCTION

Human menopausal gonadotrophins (hMGs) have been used for inducing ovulation for over 30 years, but more recently, purified preparations of urinary-derived gonadotrophins have become available which contain predominantly follicle stimulating hormone (FSH) activity. These include 'hMG 3 : 1' (Organon, Oss, The Netherlands) and Metrodin® ('pure' FSH, Urofollitrophin, Serono, Rome, Italy). The nominal ratios of FSH to luteinizing hormone (LH), as determined by *in vivo* bioassay, in the various preparations are: hMG (Humegon®; Organon or Pergonal®; Serono), 1 : 1 'hMG 3 : 1' 3 : 1 and Metrodin, > 60 : 1. When tested in an *in vitro* bioassay, using human granulosa cells, the biopotency of Metrodin has been shown to be similar to that of purified human pituitary FSH (Figure 1) and to recombinant human FSH (Org 32489, Organon)[1].

Purified preparations of gonadotrophins have theoretical advantages over the use of standard hMG but is there any evidence that there is any greater therapeutic benefit of purified FSH? This review will focus on comparative studies of the actions of 'pure' FSH, (Metrodin) and hMG, (Pergonal) in the management of patients with polycystic ovary syndrome, a disorder characterized by tonic hypersecretion of LH.

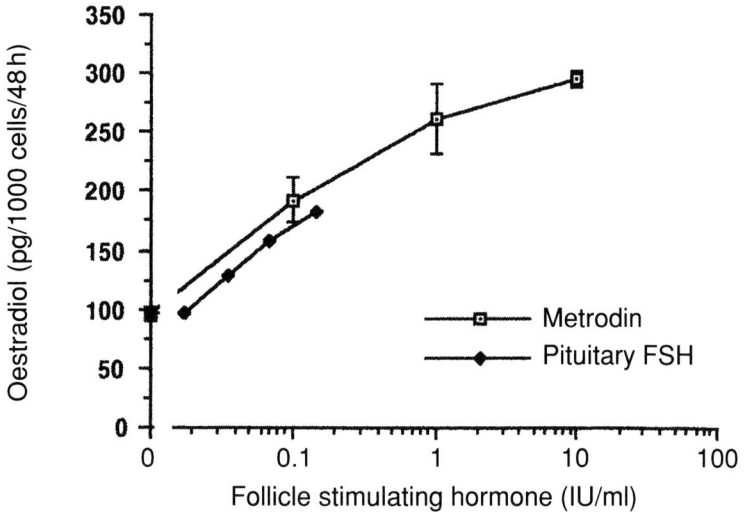

Figure 1 Comparison of effects of purified urinary follicle stimulating hormone (FSH: urofollitrophin, Metrodin®) and purified pituitary FSH on oestradiol production by human granulosa cells in culture. Cells were obtained from unstimulated human ovarian follicles after ovariectomy for benign, non-ovarian gynaecological disease. Granulosa cells were plated at a density of 50 000 cells per well in the presence of 10^{-7} mol/l testosterone as substrate. Note the similar dose-related stimulation of oestradiol by the two preparations

POTENTIAL ADVANTAGES AND DISADVANTAGES OF PURIFIED FSH

The potential advantages of an 'LH-free' FSH preparation can be categorized as either experimental or practical. From an experimental viewpoint, it is clearly useful to be able to distinguish between the individual effects of LH and FSH on the ovary. The 'two-cell, two-gonadotrophin' theory of gonadal control[2] dictates that FSH acts only on the granulosa cell compartment of the follicle, but interpretation of the effects of FSH, particularly in *in vivo* models, has, until recently, been complicated by the lack of pure FSH preparations. In this respect, recombinant FSH has obvious advantages over purified urinary FSH which has a small amount of LH 'contamination'. The experimental importance of the availability of pure FSH is well illustrated in the data from Mannaerts and de Leeuw (Chapter 1) and Hillier (Chapter 3).

The potential practical advantage of purified FSH is in the management of induction of ovulation in polycystic ovary syndrome (PCOS). This condition is a very common cause of anovulatory infertility[3,4]. Most women who require gonadotrophin treatment for ovulation induction have PCOS. The adverse effect of elevated concentrations of LH on fertility and ongoing pregnancy have been well described[5,6] and it seemed logical, therefore, to consider purified FSH for treatment of this disorder.

There is, however, a possible disadvantage in using purified FSH for induction of ovulation in women with LH deficiency. Despite the fact that complete deficiency of LH is uncommon in women with hypogonadotrophic hypogonadism, it seems unwise to consider pure FSH therapy in such cases although, to our knowledge, there have been no randomized, prospective studies in which FSH has been compared with hMG in hypogonadotrophic amenorrhoea. Functional deficiency of LH may also occur during co-treatment with gonadotrophin releasing hormone (GnRH) agonists but the circulating levels of LH during such treatment are rarely undetectable and may be sufficient to ensure that there is enough ovarian androgen production to allow a normal follicular response to FSH.

INDUCTION OF OVULATION WITH PURIFIED FSH

Efficacy of purified FSH in polycystic ovary syndrome

There have been a number of studies, over the last 10 years, which have established the effectiveness of purified urinary FSH for induction of ovulation in PCOS[7-11]. Rates of ovulation, multiple follicle development and pregnancy appear similar to those reported for hMG. Seibel and colleagues[8] claimed that the prevalence of multiple follicles and hyperstimulation was less than that observed after hMG treatment, although this study, in common with others which reported superior results with Metrodin, was not a prospective, randomized comparative one.

Low-dose purified FSH

It became clear, from these initial studies, that the use of pure FSH *per se* did not prevent the high rate of multiple follicle growth which is a particular problem of gonadotrophin therapy in PCOS. An earlier report had, however, indicated that chronic, low-dose FSH treatment was effective in inducing folliculogenesis[9]. We were interested, for this reason, in developing a low-dose method for administration of FSH. We gave a low starting dose (75 IU/day), increasing by small amounts in a stepwise fashion, using the method of Brown[12] as our model. The dose was increased (by 37.5 IU) only if there was no evidence, from ultrasound scanning, of a dominant follicle after 14 days. Subsequent increases, if necessary, were made at 7-day intervals to a maximum of 225 IU (three ampoules) per day. In the original study, Metrodin was given by intermittent subcutaneous injection using a pulsatile infusion pump[11], but subsequent data from ourselves[13] and others[14] indicated that a once-daily intramuscular dose was equally effective in limiting multiple follicle growth.

The results of these and later studies confirmed that treatment with low-dose FSH resulted in ovulation of a single follicle in more than 70% of cycles and an appropriately reduced rate of multiple pregnancy[10,15]. Are these encouraging results attributable, at least in part, to the use of purified FSH or are they simply a function of the low-dose protocol? In other words, is there any clear advantage of purified FSH over hMG? To answer this question it is necessary to consider appropriately controlled, prospective studies.

Comparative randomized studies of FSH and hMG

Larsen and colleagues studied a small number of patients to investigate, primarily, the relative safety of the two preparations given by the 'standard' (i.e. conventional dose) protocol[16]. They found no difference between Metrodin and Pergonal in terms of the number of cycles in which they observed multiple follicle development. Similar results were reported by Homburg and co-workers in a larger number of patients (46) some of whom received treatment with the GnRH agonist buserelin, before and during gonadotrophin therapy[17]. No differences were observed between the effects of FSH and hMG, irrespective of GnRH agonist treatment.

In our own study, which compares low-dose FSH with hMG, 30 clomiphene-resistant women with PCOS were randomized to treatment with either Metrodin (34 cycles in all) or Pergonal (40 cycles)[18]. The rates of ovulation, uniovulatory cycles and pregnancy were similar in the two

Table 1 Comparison of the effects of low-dose, purified follicle stimulating hormone (FSH) and human menopausal gonadotrophin (hMG) in 30 clomiphene-resistant women with polycystic ovary syndrome (data from ref. 18)

	FSH	hMG
No. patients	15	15
No. cycles	35	40
No. ovulatory cycles (%)	27 (77)	34 (85)
No. uniovulatory cycles (%)	19 (70)	22 (65)
No. pregnancies	5	5
Mean maximum daily dose (IU)	96	86
Mean dose/cycle (amps)★	19.3 (9–38.5)	14.4 (7–26)

★ranges in parentheses

groups, as were the daily and total dose requirements (Table 1). Interestingly, the mid-follicular phase suppression of serum LH concentrations which we had noted in our initial study of low-dose FSH[11] was also observed in patients who received similar doses of hMG (Figure 2). There were three early miscarriages in the Metrodin group and one in the Pergonal group but in neither our study nor in that of Homburg and colleagues[17] was the sample size large enough to comment reliably on the relative effects of the two gonadotrophins on the outcome of pregnancy.

SUMMARY

Purified, urinary-derived FSH has been used, successfully, for induction of ovulation in women with polycystic ovary syndrome. The use of a low-dose regimen greatly reduces the risk of multiple folliculogenesis and multiple pregnancy whilst maintaining efficacy. On the basis of appropriately controlled studies, however, there is no evidence that purified FSH (whether administered in low or conventional doses) is safer or more effective than hMG.

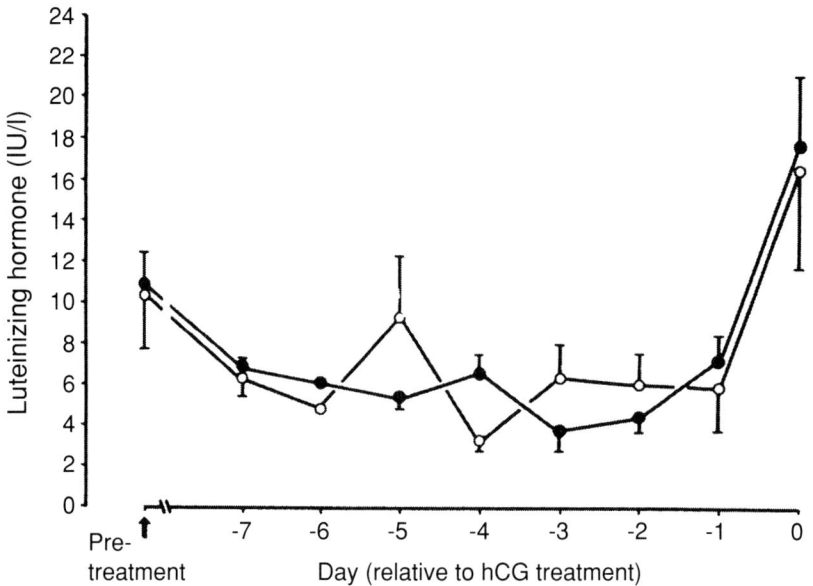

Figure 2 Serum luteinizing hormone concentrations (mean and SE) during the follicular phase of ovulatory cycles induced by human menopausal gonadotrophin (•) or follicle stimulating hormone (o) (From ref. 18, with permission)

ACKNOWLEDGEMENT

Purified pituitary follicle stimulating hormone was kindly provided by Dr S.S. Lynch, Department of Endocrinology, Birmingham and Midland Hospital for Women.

REFERENCES

1. Franks, S., Mason, H.D., de Leeuw, R. and Mannaerts, B. (1992). Stimulation of oestradiol secretion from human granulosa cells in vitro by recombinant human follicle-stimulating hormone. *J. Endocrinol.*, **132** (Suppl.), Abstr. 67
2. Armstrong, D.T. and Dorrington, J.H. (1979). Estrogen biosynthesis in the ovaries and testes. In Thomas, J.A. and Singhal, R.L. (eds.) *Regulatory*

Mechanisms Affecting Gonadal Hormone Action, vol. 2, pp. 217–58 (Baltimore: University Park Press)

3. Adams, J., Polson, D.W. and Franks, S. (1986). Prevalence of polycystic ovaries in women with anovulation and idiopathic hirsutism. *Br. Med J.*, **293**, 355–9

4. Hull, M.G. (1987). Epidemiology of infertility and polycystic ovarian disease: endocrinological and demographic studies. *Gynecol. Endocrinol.*, **1**, 235–45

5. Stanger, J.D. and Yovich, J.L. (1985). Reduced *in vitro* fertilisation of human oocytes from patients with raised basal luteinising hormone levels during the follicular phase. *Br. J. Obstet. Gynaecol.*, **92**, 385–90

6. Jacobs, H.S. and Homburg, R. (1990). The endocrinology of conception (review). *Baillières Clin. Endocrinol. Metab.*, **4**, 195–205

7. Garcea, N., Campo, S., Panetta, V., Venneri, M., Siccardi, P., Dargenio, R. and De Tomas, F. (1985). Induction of ovulation with purified urinary follicle-stimulating hormone in patients with polycystic ovarian syndrome. *Am. J. Obstet. Gynecol.*, **151**, 635–40

8. Seibel, M.M., McArdle, C., Smith, D. and Taymor, M.L. (1985). Ovulation induction in polycystic ovary syndrome with urinary follicle-stimulating hormone or human menopausal gonadotropin. *Fertil. Steril.*, **43**, 703–8

9. Kamrava, M.M., Seibel, M.M., Berger, M.J., Thompson, I. and Taymor, M.L. (1982). Reversal of persistent anovulation in polycystic ovarian disease by administration of chronic low-dose follicle-stimulating hormone. *Fertil. Steril.*, **37**, 520–3

10. Buvat, J., Buvat, H.M., Marcolin, G., Dehaene, J.L., Verbecq, P. and Renouard, O. (1989). Purified follicle-stimulating hormone in polycystic ovary syndrome: slow administration is safer and more effective. *Fertil. Steril.*, **52**, 553–9

11. Polson, D.W., Mason, H.D., Saldahna, M.B. and Franks, S. (1987). Ovulation of a single dominant follicle during treatment with low-dose pulsatile follicle stimulating hormone in women with polycystic ovary syndrome. *Clin. Endocrinol.*, **26**, 205–12

12. Brown, J.B. (1978). Pituitary control of ovarian funtion – concepts derived from gonadotrophin therapy. *Aust. NZ J. Obstet. Gynaecol.*, **18**, 47–54

13. Polson, D.W., Mason H.D., Kiddy, D.S., Winston, R.M., Margara, R. and Franks, S. (1989). Low-dose follicle-stimulating hormone in the treatment of polycystic ovary syndrome: a comparison of pulsatile subcutaneous with daily intramuscular therapy. *Br. J. Obstet. Gynaecol.*, **96**, 746–8

14. Quartero, H.W., Dixon, J.E., Westwood, O., Hicks, B. and Chapman, M.G. (1989). Ovulation induction in polycystic ovarian disease by pure

FSH (Metrodin). A comparison between chronic low-dose pulsatile administration and i.m. injections. *Hum. Reprod.*, **4**, 247–9

15. Hamilton-Fairley, D., Kiddy, D., Watson, H., Sagle, M. and Franks, S. (1991). Low-dose gonadotrophin therapy for induction of ovulation in 100 women with polycystic ovary syndrome. *Hum. Reprod.*, **6**, 1095–9

16. Larsen, T., Larsen, J.F., Schioler, V., Bostofte, E. and Felding, C. (1990). Comparison of urinary human follicle-stimulating hormone and human menopausal gonadotropin for ovarian stimulation in polycystic ovarian syndrome. *Fertil. Steril.*, **53**, 426–31

17. Homburg, R., Eshel, A., Kilborn, J., Adams, J. and Jacobs, H.S. (1990). Combined luteinizing hormone releasing hormone analogue and exogenous gonadotrophins for the treatment of infertility associated with polycystic ovaries. *Hum. Reprod.*, **5**, 32–5

18. Sagle, M.A., Hamilton-Fairley, D., Kiddy, D.S. and Franks, S. (1991). A comparative, randomized study of low-dose human menopausal gonadotropin and follicle-stimulating hormone in women with polycystic ovarian syndrome. *Fertil. Steril.*, **55**, 56–60

7

Clinical aspects of recombinant human follicle stimulating hormone

B. Mannaerts and H. Coelingh Bennink

INTRODUCTION

The expression of human follicle stimulating hormone (FSH) in Chinese hamster ovary cells transfected with both subunit genes has resulted in the synthesis of intact follicle stimulating hormone, FSH (recombinant human FSH (recFSH), Org 32489, Organon International BV). The polypeptide backbone of recFSH is identical to that of natural FSH, whereas recombinant and natural carbohydrate structures are either identical or closely related. Like natural FSH preparations, recFSH exhibits considerable charge heterogeneity and its bioactivity was confirmed by receptor displacement studies, *in vitro* bioassays and *in vivo* animal studies[1] (see also Chapter 1, this volume). Moreover, like natural FSH, recFSH induces dose-dependent increases of oestradiol and progesterone production in human granulosa cells isolated from ovaries of women with spontaneous menstrual cycles (see Chapter 6).

Both in preclinical and clinical research, the use of recFSH which is devoid of luteinizing hormone (LH) activity enables us to differentiate the effects of FSH from those of LH. In clinical conditions, the amount of endogenous and/or exogenous LH activity required to support recFSH-induced ovarian steroidogenesis is indicative for the so-called LH threshold level, which is thought to vary largely between subjects and even within one subject from cycle to cycle. The first human exposure studies of recFSH were performed in gonadotrophin–deficient male and female volunteers, mainly to assess the safety, tolerance and pharmaco-kinetic properties of recFSH. However, these studies also provided further insight into the efficacy of FSH in the presence of only minute amounts of endogenous LH. Subsequent clinical studies in infertility patients were designed to gain further knowledge about the requirements of LH during controlled superovulation. The aim of this chapter is to review some main clinical findings of recFSH which illustrate its safety and pharmacokinetic and pharmacodynamic properties.

SAFETY

Examples of therapeutic proteins produced by recombinant DNA technology and which have been shown to be safe in clinical practice are numerous and include insulin, growth hormone, erythropoietin, interferon-α and tissue plasminogen activator. A general concern with the use of recombinant proteins is their potential immunogenicity, especially if the amino-acid sequence deviates from that of the natural hormone. For recombinant glycoproteins with a peptide backbone structure identical to that of the natural hormone, such as erythropoietin or recFSH, this concern is limited to the possible minor differences in tertiary structure which are due to host cell processing. To date, many patients have been treated successfully with erythropoietin without developing specific antibodies[2]. Nevertheless, all subjects exposed to recFSH are closely monitored before and after treatment for a large number of variables, including blood biochemistry, haematology, urinalysis and the possible induction of anti-recFSH antibodies. The first human exposure studies of recFSH were performed in gonadotrophin-deficient volunteers and included multiple administrations of recFSH up to 21 days (maximal total dose 3150 IU). Monitoring of safety variables revealed no changes of clinical significance and also no drug-related adverse experiences were observed[3–5]. So far, the safety of recFSH has also been demonstrated in numerous infertility patients who have been treated with recFSH without experiencing adverse effects. First-established pregnancies[6,7], after ovarian stimulation with recFSH in patients undergoing *in vitro* fertilization and embryo transfer (IVF/ET) or ovulation induction, were reported as uneventful and proceeded normally. In October 1992, Devroey and colleagues[8] reported on the first singleton birth of a healthy baby after ovarian superovulation with recFSH.

PHARMACOKINETICS

The pharmacokinetic properties of recFSH were studied in gonadotrophin-deficient men and women in order to prevent interference by endogenous gonadotrophins. These studies included a single dose[3] and a multiple rising-dose study[4,5]. In the single-dose study a total of 300 IU recFSH was administered, while in the multiple rising-dose study, recFSH was administered once daily for a maximum of 3 weeks, i.e. 75 IU for the first 7 days, 150 IU for the next 7 days, and 225 IU during the last 7 days. Recombinant FSH was administered to all subjects by deep intramuscular injection in the upper lateral quadrant of the buttock. Serum FSH levels were assessed by means of immunoassay and *in vitro* bioassay. The immunoassay of choice was a two-site immunofluoroimmunoassay (Delfia®, Pharmacia) with a sensitivity of 0.05 IU/l in terms of International Standard (IS) 78/549. Circulating bioactive FSH was measured by means of a granulosa cell aromatase bioassay based on FSH dose-dependent increases of oestradiol secretion by rat granulosa cells *in vitro*[9]. The detection limit of this assay is 3 IU/l, in terms of IS 78/549.

Following a single dose of 300 IU recFSH, serum FSH levels were increased in all subjects after 30 min. Interestingly, recFSH was absorbed to a higher rate in men than in women, resulting in significantly higher FSH peak levels (C_{max}) in men and a significantly shorter time required to reach peak levels (T_{max}) in men, compared to women (Table 1). Consistent with these differences, the area under the serum level versus the time curve (AUC) tended to be lower in women than in men, although this difference was not significant. In the multiple-dose study, comparable absorption differences between sexes were noted, although serum FSH levels increased in accordance with the recFSH dose given, reaching steady state levels after 3–5 days in both men and women. In contrast to the above-mentioned parameters, the elimination half-lives ($T_{1/2}$) of recFSH, which were 30–40 h, were not different between men and women, either in the single dose (Table 1) or in the multiple rising-dose study. Previous pharmacokinetic studies with urinary FSH and human menopausal gonadotrophin (hMG) preparations administered via the intramuscular route have revealed comparable half-lives[10,11]. Sex differences in drug absorption and bioavailability after injection of aqueous solutions in the buttock have been described before[12] and are thought to be related to the gluteal fat thickness. In women, even

Table 1 Mean (± SD) pharmacokinetic parameters of recombinant follicle stimulating hormone (Org 32489) after one single intramuscular injection in the gluteal area of gonadotrophin-deficient men and women

	Females *n = 8*	*Males* *n = 6*★	*p*
AUC (IU/l x h)	339 ± 105	452 ± 183	NS
C_{max} (IU/l)	4.3 ± 1.7	7.4 ± 2.8	< 0.05
T_{max} (h)	27 ± 5	14 ± 8	< 0.05
$T_{1/2}$ (h)	44 ± 14	32 ± 12	NS

★ One male subject, with an extremely low body weight (42 kg), was identified as a statistical outlier and excluded from statistical evaluation

after deep injection, the drug may be placed into subcutaneous fat, leading to a less rapid absorption. In addition, differences in vascularity of the gluteus maximus are known to exist between men and women[13].

Different absorption profiles in men and women were also noted when the amounts of circulating bioactive FSH were measured. In general, immunoactive FSH levels were in good agreement with those of circulating bioactive FSH. A typical example of serum immunoactive and bioactive FSH levels of one male and one female volunteer is depicted in Figure 1.

Further evaluation of serum FSH levels after recFSH administration showed a negative correlation with body weight in both male and female gonadotrophin-deficient volunteers (Figure 2). Although body weight is known to be a determining factor for the dose and length of gonadotrophin therapy[14,15], such a relationship has not been observed previously, most probably because of interference with endogenous gonadotrophins. These findings indicate that adjustment of therapeutic doses in relation to body weight may reduce inter-subject response variability to gonadotrophin therapy.

PHARMACODYNAMICS

Clinical indications for pure (> 99.0%) recFSH are comparable to those for current urinary gonadotrophins, i.e. controlled ovarian superovulation, classical ovulation induction and male hypogonadotrophic hypogonadism.

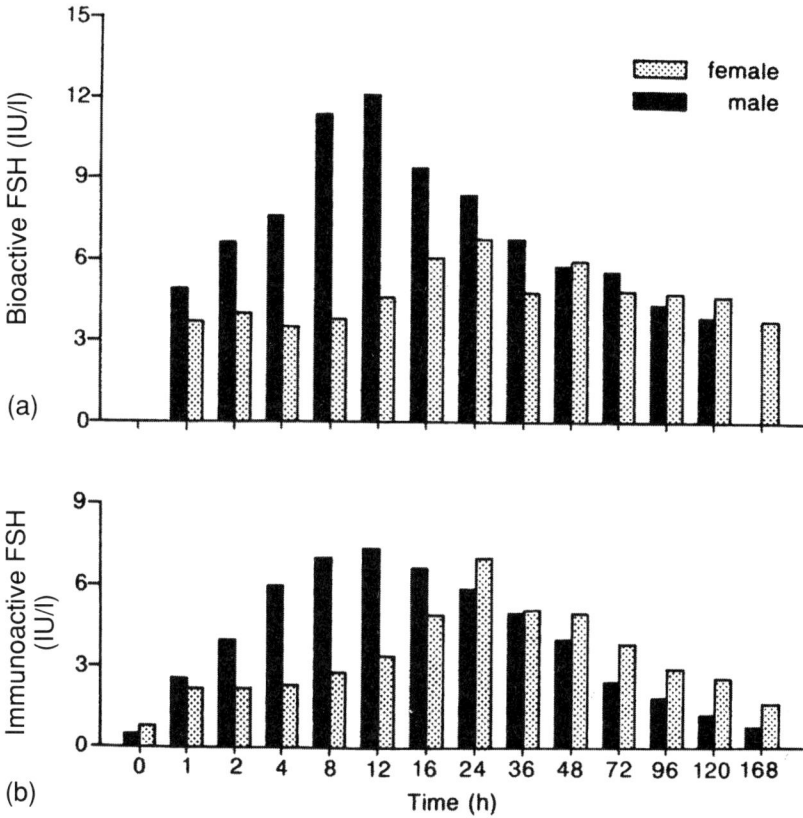

Figure 1 Serum follicle stimulating hormone (FSH) levels in terms of International Standard 78/549 measured by (a) *in vitro* granulosa cell bioassay and (b) immuno-fluorometric assay in one male and one female gonadotrophin-deficient volunteer after single intramuscular injection of 300 IU recombinant FSH (Org 32489) in the gluteal area

Current therapeutic preparations contain either urinary FSH, with minor amounts of LH activity, or human menopausal gonadotrophin (hMG), with an equal ratio of FSH and LH activity. In contrast, recFSH is devoid of LH activity. In most clinical conditions the relatively high amount of LH activity in hMG is unnecessary and might even cause premature luteinization of growing follicles. On the other hand it is recognized that some LH activity is required to support FSH-induced folliculogenesis and steroidogenesis. Studies with recFSH in hypophysectomized rats[1] have

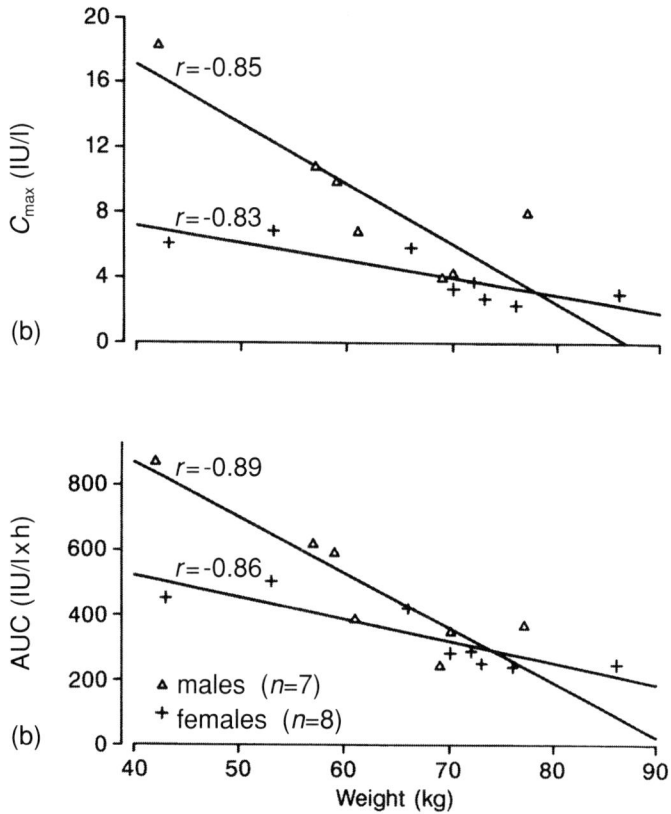

Figure 2 Correlation of (a) C_{max} and (b) area under the curve (AUC) with body weight in seven men and eight women with gonadotrophin deficiency after a single intramuscular injection of 300 IU recombinant FSH (Org 32489) in the gluteal area

demonstrated that neither recFSH nor urinary FSH were able to increase circulating levels of oestradiol, although normal follicular growth did occur. These findings supported the two-cell two-gonadotrophin theory, at least in rats, which states that both LH and FSH are required for oestrogen biosynthesis. In the case of insufficient LH activity, production of androgens in the theca cell is blocked and no aromatase substrate becomes available for conversion to oestrogens in the granulosa cell. In women with gonadotrophin deficiency, Couzinet and colleagues[16] and Shoham and co-workers[17] showed that ovulation induction with a fixed-dose regimen of urinary FSH results in lower oestrogen and androgen levels than after such treatment with hMG. Impaired oestrogen biosynthesis was also observed in the multiple rising-dose study of recFSH in gonadotrophin-deficient female volunteers. In this study daily treatment with 75–225 IU recFSH for a maximum of 21 days induced multiple follicular growth up to the preovulatory stage, whereas oestrogens and androgens remained subnormal, both in serum and follicular fluid[4,5]. Together these data suggest that the two-gonadotrophin two-cell theory is also relevant to the human, that FSH regulates granulosa cell mitogenic and steroidogenic activity differentially, and that oestrogens play a minor role in human folliculogenesis[18].

Nevertheless, oestrogens are of importance for normal endometrial function and production of cervical mucus. Therefore, the requirements of LH during controlled superovulation were evaluated by treating women undergoing IVF/ET with recFSH or with combined GnRH-agonist/recFSH regimens which are known to induce various degrees of pituitary suppression. A total of 50 women from infertile couples participated in this study and were allocated to different treatment groups, including treatment with recFSH only (Group I, $n = 9$) and treatment with recFSH in conjunction with pituitary desensitization using buserelin intranasal spray (Suprecur®), $4 \times 150\,\mu g$ per day, in a short protocol (Group II, $n = 9$) or a long protocol (Group III, $n = 11$), or using tryptorelin (Decapeptyl®) in a long protocol, giving a single dose of 3.75 mg intramuscularly (Group IV, $n = 11$) or daily subcutaneous injections of $200\,\mu g$ (Group V, $n = 10$). The case of one of the first patients in the study allocated to Group II, and successfully treated with only 9 ampoules recFSH (675 IU), was reported as the first established and ongoing pregnancy after treatment with recFSH[6].

On the day of hCG administration (10 000 IU), i.e. one day after the last recFSH injection, median serum levels of immunoactive FSH and LH were higher in patients treated without GnRH-agonist (Group I) and clearly lower in all other patients receiving combined GnRH-agonist/recFSH treatment (Table 2). Median and maximum values of immunoactive LH in each treatment group reflect pituitary suppression due to the various regimens applied, being most profound in subjects treated with tryptorelin in a long protocol (Group IV and V). On the day of hCG administration, the degree of ovarian stimulation was similar in the various treatment groups in terms of the number and sizes of follicles observed by ultrasonography, as well as the levels of serum oestradiol. However, the

Table 2 Median values of follicle stimulating hormone (FSH) and luteinizing hormone (LH) on the day of human chorionic gonadotrophin administration in patients undergoing *in vitro* fertilization/embryo transfer and treated with recombinant FSH (Group I) or combined GnRH–agonist/recombinant FSH regimens (Groups II–V). Ranges are given in parentheses

	Treatment groups				
	I	*II*	*III*	*IV*	*V*
No. patients	9	9	11	11	10
FSH (IU/l)	21 (14–26)	13 (4–17)	15 (0.3–24)	17 (9.1–27)	17 (10–30)
LH (IU/l)	5.1 (1.2–20)	2.3 (0.5–20)	1.3 (0.5–7.1)	1.2 (0.8–3.5)	1.6 (0.5–2.7)

number of treatment days, the total number of ampoules and the number of ampoules per day required to establish ovarian stimulation increased with the apparent degree of pituitary suppression (Table 3). For example, the total amount of recFSH required in Group IV, treated with tryptorelin intramuscular depot, was twice as high as that required in control Group I, treated with recFSH only. Successful superovulation and pregnancies were established, regardless of the GnRH–regimen applied.

Together, these data suggest that the remaining endogenous LH activity after pituitary suppression with GnRH–agonists is sufficient to support recFSH treatment for controlled ovarian superovulation. However, the amount of LH activity might influence ovarian responsiveness and, thus, the amount of recFSH required for successful ovarian stimulation. Since the number of observations in this study was limited and the intersubject variability large, additional clinical research will be required to confirm our preliminary findings. Further clinical research is also required to establish the long-term safety and efficacy of recFSH, with special emphasis on the adverse aspects of current urinary gonadotrophin preparations, i.e. multiple pregnancy and ovarian hyperstimulation.

Table 3 Stimulation characteristics of patients undergoing *in vitro* fertilization/ embryo transfer and treated with recombinant follicle stimulating hormone (recFSH) (Group I) or combined GnRH–agonist/recFSH regimens (group II to V). Values represent the mean ± SD

	Treatment groups				
	I	*II*	*III*	*IV*	*V*
No. patients	9	9	11	11	10
Treatment days	7.3 ± 1.3	12.1 ± 3.5	12.6 ± 2.3	14.5 ± 2.2	13.0 ± 2.3
No. ampoules	21.3 ± 4.5	24.7 ± 13.6	35.0 ± 10.3	40.5 ± 16.7	32.6 ± 10.0
No. ampoules/day	2.9 ± 0.2	1.9 ± 0.6	2.7 ± 0.4	2.7 ± 0.8	2.5 ± 0.5

SUMMARY

In comparison with commercially available urinary gonadotrophin preparations, recombinant human follicle stimulating hormone (recFSH, Org 32489) has a very high purity (> 99.0%) and lacks luteinizing hormone activity. These properties of recFSH prompted further clinical research examining its safety, tolerance, pharmacokinetic and pharmacodynamic properties. Single-dose and multiple rising-dose studies of recFSH in gonadotrophin-deficient male and female volunteers revealed that recFSH is safe and well-tolerated, and no drug-related adverse experiences were noted following once-daily administration for up to 3 weeks. Pharmacokinetic evaluation of serum FSH concentrations indicated that after intramuscular administration of recFSH in the buttock, a slower rate of absorption was noted in women than in men, whereas their elimination half-lives are comparable and in agreement with those reported for natural FSH. Multiple rising-dose administrations of recFSH resulted in dose-related increases of circulating FSH. In gonadotrophin-deficient women, daily administration of recFSH induced follicular growth up to the preovulatory stage whereas oestrogen and androgen levels in serum and follicular fluid remained extremely low. These observations indicate that the two-cell two-gonadotrophin theory, holding that both FSH and LH are required for oestrogen biosynthesis, is also relevant to the human. In patients receiving combined GnRH-agonist/recFSH treatment, the amount of remaining endogenous LH seems to determine the number of treatment days and total amount of recFSH required for successful superovulation. It is suggested that the amount of LH activity influences the initial ovarian responsiveness. Further clinical research on recFSH is required to establish the long-term safety and efficacy of combined GnRH-agonist/recFSH treatment in female infertility.

ACKNOWLEDGEMENTS

Clinical investigators in this Org 32489 research project were: Prof. H. Jacobs (Middlesex Hospital, London, United Kingdom), Dr. Z. Shoham (Kaplan Hospital, Rehovot, Israel), Dr. B. Fauser (Dijkzigt University Hospital, Rotterdam, The Netherlands), Dr. D. Schoot (Westeinde Hospital, The Hague, The Netherlands), Prof. Ph. Bouchard (Hospital Bicètre, Le Kremlin Bicètre, France), Dr. J. Harlin (Karolinska Hospital, Stockholm, Sweden), Prof. P. Devroey and Prof. A. van Steirteghem (Free University of Brussels, Brussels, Belgium). Serum bioactive FSH levels were assessed by Dr. K. Dahl (Veterans Administration Medical Centre, Seattle, Washington, USA).

REFERENCES

1. Mannaerts, B., De Leeuw, R., Geelen, J., Van Ravenstein, A., Van Wezenbeek, P., Schuurs, A. and Kloosterboer, L. (1991). Comparative *in vitro* and *in vivo* studies on the biological properties of recombinant human follicle stimulating hormone. *Endocrinology*, **129**, 2623–30
2. Watson, A.J. (1989). Adverse effects of therapy for the correction of anemia in hemodialysis patients. *Semin. Nephrol.*, **9**, 30–4
3. Mannaerts, B., Shoham, Z., Schoot, B., Bouchard, Ph., Harlin, J., Fauser, B., Jacobs, H., Rombout, F. and Coelingh Bennink, H. (1993). Single dose pharmacokinetics and pharmacodynamics of recombinant human follicle-stimulating hormone (Org 32489) in gonadotropin-deficient volunteers. *Fertil. Steril.*, **59**, 108–14
4. Schoot, B.C., Coelingh Bennink, H.J., Mannaerts, B.M., Lamberts, S.W., Bouchard, P. and Fauser, B.C. (1992). Human recombinant follicle-stimulating hormone induces growth of preovulatory follicles without concomitant increase in androgen and estrogen biosynthesis in a woman with isolated gonadotropin deficiency. *J. Clin. Endocrinol. Metab.*, **74**, 1471–3

5. Shoham, Z., Mannaerts, B., Insler, V. and Coelingh Bennink, H. (1993). Induction of follicular growth using recombinant human follicle-stimulating hormone in two volunteer women with hypogonadotropic hypogonadism. *Fertil. Steril.*, in press

6. Devroey, P., Van Steirteghem, A., Mannaerts, B. and Coelingh Bennink, H. (1992). Successful *in-vitro* fertilization and embryo transfer after treatment with recombinant human FSH. *Lancet*, **339**, 1170-1

7. Donderwinkel, P.F.J., Schoot, D.C., Coelingh Bennink, H.J.T. and Fauser, B.C.J.M. (1992). Pregnancy after induction of ovulation with recombinant human FSH in polycystic ovary syndrome. *Lancet*, **340**, 983

8. Devroey, P., Van Steirteghem, A., Mannaerts, B., Coelingh Bennink, H. (1992). First singleton term birth after ovarian superovulation with recombinant human follicle stimulating hormone (Org 32489). *Lancet*, **340**, 1108

9. Dahl, K.D. and Hsueh, A.J.W. (1989). Granulosa cell aromatase bioassay for follicle-stimulating hormone. *Meth. Enzymol.*, **169**, 414–23

10. Diczfalusy, E. and Harlin, J. (1988). Clinical pharmacological studies on human menopausal gonadotrophin. *Hum. Reprod.*, **3**, 21–7

11. Mizunuma, H., Takagi, T., Honjyo, S., Ibuki, Y. and Igarashi, M. (1990). Clinical pharmacodynamics of urinary follicle-stimulating hormone and its application for pharmacokinetic simulation program. *Fertil. Steril.*, **53**, 440–5

12. Zuidema, J., Pieters, F.A. and Duchateau, G.S. (1988). Release and absorption rate aspects of intramuscular injected pharmaceuticals. *Int. J. Pharmaceutics*, **47**, 1–12

13. Vukovich, R.A., Brannick, L.J., Sugerman, A.A. and Neiss, E.S. (1976). Sex differences in the intramuscular absorption and bioavailability of cephradine. *Clin. Pharmacol. Ther.*, **18**, 215–20

14. Hedon, B., Bringer, J., Fries, N., Thomas, G., Pelliccia, G., Bachelard, B., Benos, P., Arnal, F., Humeau, C., Nares, P., Laffargue, F. and Viala, J.L. (1991). Influence of weight on the ovarian response to the stimulation of ovulation for *in vitro* fertilization. *Contracept. Fertil. Sex*, **19**, 1037–41

15. Chong, A.P., Rafael, R.W. and Forte, C.C. (1986). Influence of weight in the induction of ovulation with human menopausal gonadotropin and human chorionic gonadotropin. *Fertil. Steril.*, **46**, 599–603

16. Couzinet, B., Lestrat, N., Brailly, S., Forest, M. and Schaison, G. (1988). Stimulation of ovarian follicular maturation with pure follicle stimulating hormone in women with gonadotropin deficiency. *J. Clin. Endocrinol. Metab.*, **66**, 552–6

17. Shoham, Z., Balen, A., Patel, A. and Jacobs, H.S. (1991). Results of ovulation induction using human menopausal gonadotropin or purified

follicle-stimulating hormone in hypogonadotropic hypogonadism patients. *Fertil. Steril.*, **56**, 1048–53

18. Chappel, S.C. and Howles, C. (1991). Reevaluation of the roles of luteinizing hormone and follicle-stimulating hormone in the ovulatory process. *Hum. Reprod.*, **9**, 1206–12

8

Management of ovulatory disorders with pulsatile gonadotrophin releasing hormone

M. Filicori, G. Cognigni, L. Michelacci, P. Dellai, M. Sambataro and F. Carbone

INTRODUCTION

Ovulation induction is a commonly used therapeutic approach in the management of infertility. Several compounds can be employed for this treatment; each is characterized by different endocrine and clinical properties that permit a rationale for the choice of the optimal method.

Clomiphene citrate is an effective and widely used drug[1]; this antioestrogen appears to be relatively efficient in inducing ovulation, particularly in the polycystic ovary syndrome (PCOS) and in non-severe forms of menstrual disorders. Clomiphene is ineffective in progestogen-negative amenorrhoea and the pregnancy rates obtained with this method are relatively low compared to the apparent ovulatory rates. It is therefore conceivable that inadequate ovulation or follicular luteinization occur in some patients or that other problems (e.g. cervical mucus antioestrogenic actions) prevent conception. However, when used alone (i.e. without human chorionic gonadotrophin (hCG)) clomiphene is associated with a limited risk of complications (such as ovarian hyperstimulation and multiple pregnancy) and does not require endocrine monitoring. Therefore, the use of this medication is indicated as a primary level approach when ease of use rather than effectiveness is preferred.

The gonadotrophins hMG (human menopausal gonadotrophins) and purified FSH (follicle stimulating hormone) are drugs which represent an extremely effective approach to ovulation induction because of their direct stimulatory action on the ovary[2]. However, for the same reason, gonadotrophins are often associated with excessive ovarian stimulation and its complications. Endocrine monitoring (daily oestrogen determinations) and pelvic ultrasonography should never be withheld during hMG stimulation. However, controlled ovarian hyperstimulation is a prerequisite rather than a side-effect when ovulation induction is followed by one of the many techniques of assisted reproduction. More recently, gonadotrophin ovulation induction has been supplemented by gonadotrophin releasing hormone (GnRH) analogue or growth hormone co-administration. Both these drugs appear to improve the ovulation induction response, albeit through different mechanisms.

The finding in the late 70s that intermittent administration of GnRH is optimally effective in stimulating pituitary and gonadal function prompted the use of this medication for clinical ovulation induction. Extensive experience in the use of pulsatile GnRH has been gained in the last decade[3]. The remainder of this chapter will be devoted to the technical and clinical aspects of ovulation induction with pulsatile GnRH.

REGIMENS OF GnRH ADMINISTRATION

Route and dose

Pulsatile GnRH can be administered both via the intravenous and the subcutaneous route. Although there is a theoretical possibility of severe septical complications when GnRH is administered intravenously, this route has so far proven remarkably safe. An investigation of this aspect showed that[4] few positive blood cultures occurred with a limited relationship to actual clinical problems. Changing the intravenous site is sometimes needed when phlebitis develops and anti-inflammatory and/or antibiotic therapy is only rarely required.

The major advantage of intravenous GnRH administration is the closer resemblance of the dynamics of exogenous GnRH pulses to endogenous pulsatile GnRH secretion that is achieved[5]. The subcutaneous route results in blunted GnRH pulses and reduced hormone absorption. Thus, larger GnRH doses may be required if the subcutaneous route is preferred. Clinical results appear to be improved by the use of intravenous GnRH.

A wide range of GnRH dosages have been employed for ovulation induction in females, ranging from 1 to 40 µg/bolus or higher[3]. When an optimal GnRH interpulse interval of 60 or 90 min is employed (see below), adequate pituitary and ovarian stimulation can be achieved with low-dose GnRH. Although the GnRH dose currently suggested for ovulation induction is 5 µg/bolus (Figure 1), almost perfect restoration of the menstrual cycle endocrine dynamics can be obtained with a dose of 2.0–2.5 µg/bolus[6,7]. Low-dose pulsatile GnRH administration is preferable, because of the higher incidence of multiple pregnancy associated with larger GnRH dosages[8]. If a non-physiological pulse frequency of 120 min is adopted, higher GnRH dosages may be required to achieve comparable responses[9].

Frequency of administration

In the follicular phase of the normal menstrual cycle the spontaneous LH pulses induced by endogenous GnRH occur at a roughly circhoral frequency[10]. Most ovulatory disorders are associated with derangements of hypothalamic-hypophyseal pulses that may occur too rapidly as in PCOS[11,12] or too slowly[13]. Even modest slowing of LH peaks from 1 per hour to 1 every 2 hours is not compatible with normal ovulatory function[13]. Thus, a precise pattern of exogenous pulsatile GnRH administration is critical for an optimal outcome of this therapy. We demonstrated previously[9] that GnRH administration every 120 min is less effective than every 60 min to induce ovulation in profound hypogonadotrophic hypogonadism, even when a greater GnRH dose/bolus is employed; a slow GnRH frequency blunts the midcycle LH surge and is associated with reduced ovulatory rates. More recently we demonstrated that a 90 min interval may be less than optimal for ovulation induction[14]. Thus, administration of GnRH at 60 min intervals should be employed, whenever possible.

Specific drug regimens

Although simple pulsatile GnRH administration is usually adequate for ovulation induction in most patients with hypogonadotrophic hypo-gonadism (Figure 1) this does not hold true for other disorders. Patients

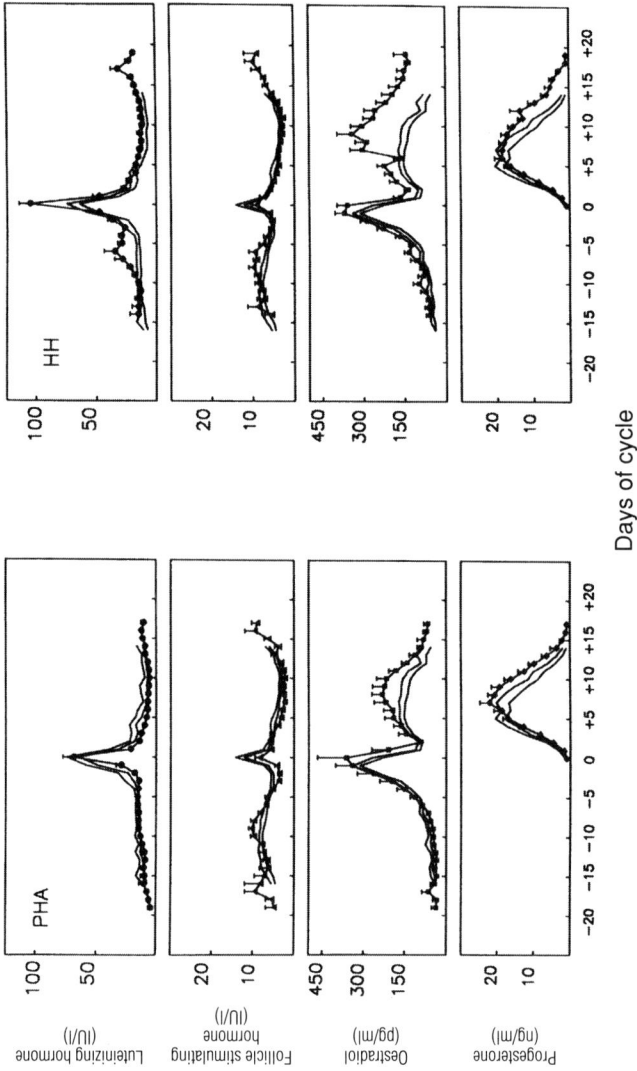

Figure 1 Daily gonadotrophin and gonadal steroid levels (mean ± SE) during ovulation induction with pulsatile GnRH (5 μg, every 60 min) in different hypogonadotrophic disorders. Left panel: patients with primary hypogonadotrophic amenorrhoea (PHA); right panel: other forms of hypogonadotrophic hypogonadism (HH). The area between the continuous lines denotes values of normal controls. Note the almost perfect overlap of PHA values with normal cases. Reproduced, with permission, from ref. 6

with PCOS are generally more resistant to this form of ovulation induction and may require a specific management. In 1988 we demonstrated that pituitary–ovarian suppression with a GnRH agonist immediately before pulsatile GnRH renders PCOS patients more responsive to this form of ovulation induction[11]. GnRH analogues reduce LH and androgen secretion and create an ovarian environment more favourable for follicular development, as excessive intra-ovarian androgen levels enhance follicular atresia[15]. When this combined regimen of GnRH analogue/pulsatile GnRH was applied to a larger series of PCOS subjects (Figure 2) an ovulatory rate of 76% was obtained with pregnancy rates of 28%[6]. Unfortunately, the endocrine improvement induced by GnRH analogues is transitory and maintained only in the first post-analogue cycle[16]. We also found that this drug regimen is effective in reducing the excessive pituitary–ovarian stimulation occasionally encountered also in hypogo-nadotrophic patients[6], and that may be responsible for multiple pregnancy resulting from pulsatile GnRH ovulation induction in these patients.

CLINICAL RESULTS

A general agreement exists on the high efficacy of pulsatile GnRH ovulation induction in hypogonadotrophic hypogonadism[3]. An ovulatory rate of about 90% of treatment cycles can be obtained and pregnancy rates are between 20 and 30% per cycle, i.e. comparable to the estimated fecundity level of normal women. No cases of ovarian hyperstimulation have ever been reported with pulsatile GnRH, thus rendering endocrine monitoring superfluous. Multiple pregnancy is possible with pulsatile GnRH and its risk is around 10% of treatment cycles, which is markedly lower than with the use of gonadotrophins.

Treatment outcome is far poorer in PCOS patients[16,17] (*ca.* 40% ovulatory cycles and < 10% pregnancy rates). However, as previously discussed, GnRH analogue pretreatment (buserelin 300 μg, subcutaneously every 12 h for 6 weeks) improves the endocrine and clinical outcome of these patients (Figure 2)[11]. An additional problem in PCOS patients is weight, as success rates are further reduced in obese PCOS subjects[6]. Nevertheless, when non-obese PCOS subjects are pretreated with a GnRH analogue, ovulatory rates improve drastically and are comparable to those of hypogonadotrophic hypogonadism.

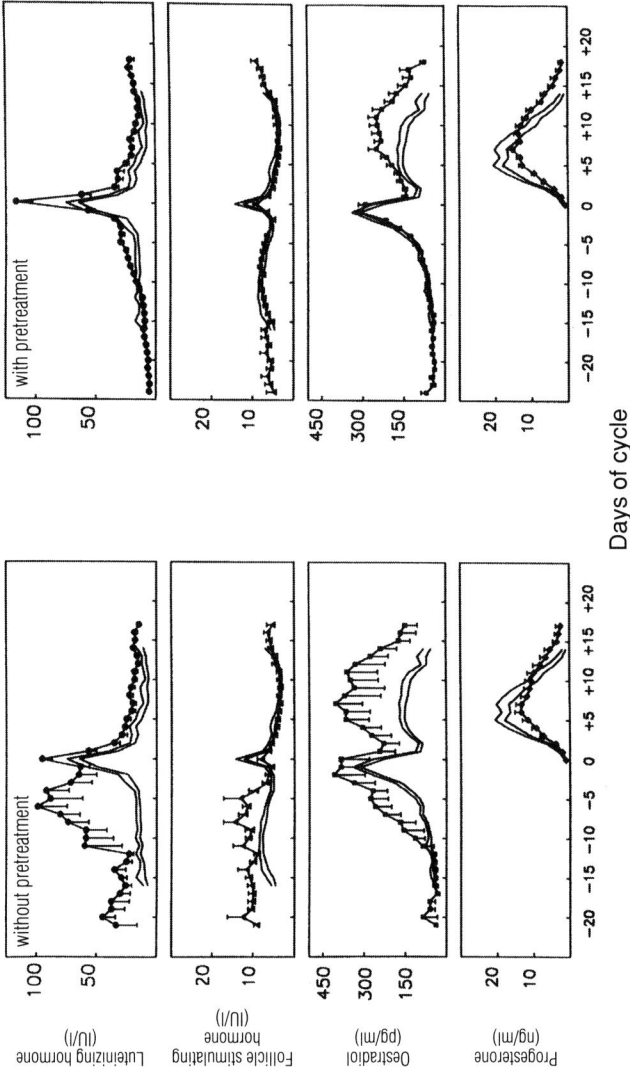

Figure 2 Daily gonadotrophin and gonadal steroid levels (mean ± SE) during ovulation induction with pulsatile GnRH (5 μg, every 60 min) in PCOS patients. Cycles depicted in left panels were not pretreated while patients on the right had a course of GnRH analogue (buserelin 300 μg, subcutaneously, twice a day for 6 weeks) before ovulation induction (see text for details). Note the almost complete endocrine normalization of the treated cycles. Reproduced, with permission, from ref. 6

Another complication of ovulation induction regimens is spontaneous abortion. Whilst miscarriage in hypogonadotrophic patients treated with pulsatile GnRH is comparable to that occurring in spontaneous pregnancy (*ca.* 10%), a higher incidence of this complication (*ca.* 40%) should be expected in PCOS[6,17]. Unfortunately, the pathogenetic mechanism of abortion in these patients is still unknown and no therapeutic option is presently available; GnRH analogue pretreatment will not reduce spontaneous abortion rates in PCOS[6].

CONCLUSIONS

Pulsatile GnRH is a safe and effective method for ovulation induction in a wide variety of ovulatory disorders. The use of pulsatile GnRH avoids complex and expensive endocrine monitoring, eliminates the risk of ovarian hyperstimulation, and reduces the risk of multiple pregnancy. Therefore, pulsatile GnRH should be preferred to gonadotrophins and be the treatment of first choice for cases of anovulatory infertility outside assisted reproduction protocols, who are clomiphene-resistant or non-responders.

ACKNOWLEDGEMENTS

We thank Ms Silvia Arsento for outstanding secretarial assistance. This work was supported in part by grants from the National Research Council of Italy, ASTO Bologna, and the Regional Government of Emilia-Romagna.

REFERENCES

1. Adashi, E.Y. (1984). Clomiphene citrate mechanism(s) and site(s) of action – a hypothesis revisited. *Fertil. Steril.*, **42**, 331–40
2. Oelsner, G., Serr, D.M., Mashiach, S., Blankestein, J., Snyder, M. and Lunenfeld, B. (1978). The study of induction of ovulation with menotropins: analysis of results of 1987 treatment cycles. *Fertil. Steril.*, **30**, 538–43
3. Filicori, M., Flamigni, C., Meriggiola, M.C., Cognigni, G., Valdiserri, A., Ferrari, P. and Campaniello, E. (1991). Ovulation induction with pulsatile

gonadotropin-releasing hormone: technical modalities and clinical perspectives. *Fertil. Steril.*, **56**, 1–13

4. Hopkins, C.C., Santoro, N.F., Hall, J.E., Martin, K.A., Wierman, M.E., Waldstreicher, J., Filicori, M. and Crowley, W.F. Jr (1989). Closed intravenous administration of gonadotropin-releasing hormone (GnRH): safety of long duration of peripheral intravenous catheterization. *Obstet. Gynecol.*, **74**, 267–70

5. Couzinet, B., Lahlou, N., Lestrat, N., Bouchard, P., Roger, M. and Schaison, G. (1986). Pulsatile luteinizing hormone-releasing hormone treatment for induction of ovulation. Radioimmunoassay of plasma LHRH and comparative study of subcutaneous versus intravenous routes of administration. *J. Endocrinol. Invest.*, **9**, 103–8

6. Filicori, M., Flamigni, C., Meriggiola, M.C., Ferrari, P., Michelacci, L., Campaniello, E., Valdiserri, A. and Cognigni, G. (1991). Endocrine response determines the clinical outcome of pulsatile GnRH ovulation induction in different ovulatory disorders. *J. Clin. Endocrinol. Metab.*, **72**, 965–72

7. Santoro, N., Wierman, M.E., Filicori, M., Waldstrecher, J. and Crowley, W.F. Jr (1986). Intravenous administration of pulsatile gonadotropin-releasing hormone in hypothalamic amenorrhea: effects of dosage. *J. Clin. Endocrinol. Metab.*, **62**, 109–16

8. Braat, D.D.M., Ayalon, D., Blunt, S.M., Bogchelman, D., Coelingh-Bennink, H.J.T., Handelsman, D.J., Heineman, M.J., Lappohn, R.E., Lorijn, R.H.W., Rolland, R., Willemsen, W.M.P. and Shoemaker, J. (1989). Pregnancy outcome in luteinizing hormone-releasing hormone induced cycles: a multicentre study. *Gynecol. Endocrinol.*, **3**, 35–40

9. Filicori, M., Flamigni, C., Campaniello, E., Ferrari, P., Meriggiola, M.C., Michelacci, L., Pareschi, A. and Valdiserri, A. (1989). Evidence for a specific role of GnRH pulse frequency in the control of the human menstrual cycle. *Am. .J. Physiol.*, **257**, E930–6

10. Filicori, M., Santoro, N., Merriam, G.R. and Crowley, W.F. (1986). Characterization of the physiological pattern of episodic gonadotropin secretion throughout the human menstrual cycle. *J. Clin. Endocrinol. Metab.*, **62**, 1136–44

11. Filicori, M., Campaniello, E., Michelacci, L., Pareschi, A., Ferrari, P., Bolelli, G. and Flamigni, C. (1988). Gonadotropin-releasing hormone (GnRH) analog suppression renders polycystic ovary disease patients more susceptible to ovulation induction with pulsatile GnRH. *J. Clin. Endocrinol. Metab.*, **66**, 327–33

12. Waldstreicher, J., Santoro, N., Hall, J.E., Filicori, M. and Crowley, W.F. (1988). Hyperfunction of the hypothalamic–pituitary axis in women with

polycystic ovarian disease: indirect evidence for partial gonadotrophin desensitization. *J. Clin. Endocrinol. Metab.*, **66**, 165–172

13. Reame, N.E., Sauder, S.E., Case, G.D., Kelch, R.P. and Marshall, J.C. (1985). Pulsatile gonadotropin secretion in women with hypothalamic amenorrhea: evidence that reduced frequency of gonadotropin-releasing hormone secretion is the mechanism of persistent anovulation. *J. Clin. Endocrinol. Metab.*, **61**, 851–8

14. Meriggiola, M.C., Flamigni, C., Valdiserri, A., Ferrari, P., Michelacci, L., Campaniello, E. and Filicori, M. (1990). Lack of discernible effects of a constantly rapid pulsatile GnRH frequency in the luteal phase (LP) upon gonadal function of subsequent menstrual cycles. Presented at the *72nd Annual Meeting of the Endocrine Society*, Atlanta, June

15. Louvet, J.P., Harman, S.M., Schreiber, J.R. and Ross, G.T. (1975). Evidence for a role of androgens in follicular maturation. *Endocrinology*, **97**, 366-72

16. Filicori, M., Flamigni, C., Campaniello, E., Valdiserri, A., Ferrari, P., Meriggiola, M.C., Michelacci, L. and Pareschi, A. (1989). The abnormal response of polycystic ovarian disease patients to exogenous pulsatile gonadotropin-releasing hormone: characterization and management. *J. Clin. Endocrinol. Metab.*, **69**, 825–31

17. Homburg, R., Eshel, A., Armar, N.A., Tucker, M., Mason, P.W., Adams, J., Kilborn, J., Sutherland, I.A. and Jacobs, H.S. (1989). One hundred pregnancies after treatment with pulsatile luteinising hormone-releasing hormone to induce ovulation. *Br. Med. J.*, **298**, 809–812

9

Growth hormone and ovarian stimulation

H.S. Jacobs

INTRODUCTION

The first evidence to suggest a role for growth hormone (GH) in the ovulatory process was the observation that lowering GH levels in female rats led to delayed puberty and decreased ovarian steroidogenesis in response to stimulation by gonadotrophins[1]. Moreover, in children with Laron type dwarfism[2], puberty is delayed and prolonged, while in hypophysectomized animals gonadal maturation can be induced by the administration of GH[3].

The second line of evidence is more deductive. On reviewing the endocrine events of the human ovulation cycle, during the follicular phase, one notes a striking contrast between the explosive increase in oestradiol secretion, reflecting follicular growth, and the very modest increase in the concentrations of follicle stimulating hormone (FSH). This difference suggests the possibility that ovarian paracrine factors may amplify the effect of FSH on follicle growth. Adashi and colleagues[4] summarized information on the paracrine control of follicular activity: it seems that insulin-like growth factors (IGFs) are particularly relevant. In murine systems their synergistic effects with FSH on granulosa differentiation have been demonstrated through stimulation of aromatase activity, induction of LH receptors, synthesis of progestins and proteoglycans, and cyclic adenosine monophosphate accumulation[4]. While there is evidence that IGF-1 is synthesized by mature rat ovarian follicles in which RNA

97

transcripts have been identified[5,6], evidence that IGF-1 is synthesized in human granulosa is less certain. Growth factors, however, may be synthesized in theca cells.

At the very time that these *in vitro* studies were published, we had found that a number of our patients undergoing induction of ovulation were resistant to treatment with gonadotrophins. We hypothesized that by co-treatment with GH we might increase endogenous IGF-1 concentrations and thereby sensitize the ovaries to stimulation by gonadotrophins. We thought that the putative action of GH would operate through a paracrine mechanism – e.g. that treatment with GH would increase production of IGF-1 within the ovarian follicle itself. While the first of our clinical studies were open we soon moved to randomized, placebo-controlled studies for induction of ovulation for *in vivo* and *in vitro* fertilization. The latter study was performed in order to gain access to follicular fluid from GH-treated cycles, but we also used the experimental setup to determine whether, for the same dose of gonadotrophins, women who had been co-treated with GH would yield more follicles and more oocytes than women who were co-treated with placebo.

CONTROLLED CLINICAL TRIAL OF CO-TREATMENT WITH HUMAN GROWTH HORMONE AND GONADOTROPHINS FOR INDUCTION OF OVULATION IN ANOVULATORY PATIENTS

We studied 16 patients with amenorrhoea who required human menopausal gonadotrophins (hMG) for induction of ovulation and who previously had been treated for at least one cycle in which a minimum of 25 hMG ampoules were needed. Five patients had had pituitary surgery for macroprolactinoma, pituitary-dependent Cushing's disease, chromophobe adenoma and craniopharyngioma. Five patients suffered from hypogonadotrophic hypogonadism, one had Kallmann's syndrome and five had ultrasound-diagnosed polycystic ovaries (PCOs).

Patients were assigned to co-treatment with either GH or placebo using a blinded randomization procedure. Eight patients received GH and hMG and eight received placebo with hMG. The assignment code was broken at the end of each treatment cycle. Those who received GH were considered to have completed the study, whereas those who had received placebo were then offered co-treatment with GH in a subsequent cycle.

In all treatment cycles hMG was administered according to the individually adjusted dose scheme, starting with one ampoule per day for 5 days and increasing the dose by one ampoule per day until a sufficient ovarian response was obtained, as detected by ultrasound. The 'daily effective dose' was then continued until one or more follicles attained a size of > 17 mm in diameter. In order to induce ovulation, human chorionic gonadotrophin, (hCG) 10 000 IU, was administered. Growth hormone, 24 IU, by intramuscular injection, was administered on alternate days to a total dose of 144 IU, or until the day of hCG administration, starting on the first day of the hMG therapy. Patients who needed treatment for more than 12 days continued with hMG but not with GH. Placebo was administered in the same way as the GH.

The mean ages of patients in the GH and the placebo group were 31.9 (range 23–40 years) and 30.1 (range 26–34 years), respectively. The mean ($\pm$ SD) concentrations of serum luteinizing hormone (LH), FSH, and oestradiol concentrations were 2.7 ± 1.7 IU/l, 1.6 ± 1.5 IU/l and 62.5 ± 35 pmol/l, respectively. The various clinical diagnoses were equally distributed between the two treatment groups and, in particular, there were two hypophysectomized patients in the placebo group and three in the GH-treated group.

Results

Three patients conceived in their first study cycle: one in the placebo and two in the GH group. During the GH treatment, the number of ampoules of hMG required to induce follicular development was reduced ($p = 0.008$), the duration of treatment was shorter ($p = 0.011$) and the 'daily effective dose' was lower, ($p = 0.035$) compared with the placebo cycles (Table 1). The number of follicles that were induced to a diameter of > 17 mm was marginally higher in the placebo (2.4 ± 1.1) than in the GH group (1.4 ± 0.7), ($p = 0.05$). The number of cohort follicles (14–16 mm in diameter) was not significantly different in the placebo (2.9 ± 2.2) and GH groups (2.4 ± 1.4; $p = 0.58$). The range of serum oestradiol concentrations was 850–7340 pmol/l and there were no significant differences between the groups. All placebo cycles, and all but one of the GH-treated cycles, were ovulatory, as detected by ultrasound scanning and by midluteal serum progesterone concentrations. Serum

Table 1 Comparison of the number of ampoules, days of treatment and daily effective dose of human menopausal gonadotrophin (hMG) in placebo and GH (growth hormone) groups (values are mean ± SD)

Variable	Group	No. cycles	Prestudy cycle	Treatment cycle
No. hMG	placebo	8	43.6 ± 13.1	42.5 ± 13.1
ampoules	GH	8	34.4 ± 7.8	24.5 ± 9.7
			$p = 0.11$	$p = 0.008$
Days of	placebo	8	19.3 ± 3.7	18.5 ± 3.8
treatment	GH	8	17.3 ± 2.3	13.4 ± 3.2
			$p = 0.22$	$p = 0.011$
Daily effective	placebo	8	3.13 ± 0.83	3.38 ± 0.74
hMG dose	GH	8	2.88 ± 0.64	2.50 ± 0.76
			$p = 0.51$	$p = 0.035$

IGF-1 concentrations rose during GH treatment and peaked between the second and third injection at a mean of more than twice the upper limit of normal, falling back into the normal range within one week of the last GH injection. There was no change in serum IGF-2 concentrations. Neither of the growth factor concentrations changed significantly during placebo cycles.

A PROSPECTIVE, RANDOMIZED, DOUBLE-BLIND, PLACEBO-CONTROLLED TRIAL IN PATIENTS UNDER-GOING IVF-ET FOLLOWING PITUITARY SUPPRESSION

Study design and patients

Patients undergoing *in vitro* fertilization–embryo transfer (IVF-ET) at Hallam Medical Centre, London, between January and December 1989, were studied. We recruited subjects who were less than 38 years of age and who had undergone one or more IVF-ET cycles in which ovarian stimulation had been carried out using the combined regimen of luteinizing hormone releasing hormone (LHRH) agonist and hMG, and in which the response had been considered suboptimal (defined as one in which fewer than six oocytes were collected from which fewer than four embryos developed).

A total of 25 patients were recruited, all of whom had had a pretreatment ultrasound scan to determine ovarian morphology. According to the ovarian scan the patients were divided into those with normal ovaries and those with ultrasound-diagnosed PCOs. The distinction was based on the criteria of Adams and colleagues[7]. Patients in both groups were then randomized to receive GH or placebo, in addition of course to their standard treatment for IVF. Thirteen patients were allocated to receive GH (24 units per injection given intramuscularly), and 12 to receive placebo injections, starting on the first day of hMG treatment. The GH or placebo was given on alternate days until the day of administration of hCG, or for a maximum period of 2 weeks. At the completion of the cycle of treatment the assignment code was broken. Those who had received GH were considered to have completed the study, and those who had received placebo entered an open study in which they received GH, so that they were not deprived of a potentially beneficial therapy. The results of the open study are not included here. In all cases, an interval of 2 months was allowed to elapse between cycles of treatment.

All patients had had at least one previous IVF-ET attempt using pituitary gonadotrophin suppression (buserelin, Suprefact®, Hoechst, Hounslow, U. K.) 200 µg subcutaneously. In all treatment cycles, the analogue was administered daily from the first day of the menstrual period, for a minimum of 14 days. When ovarian suppression was confirmed (serum oestradiol concentrations < 150 pmol/l), treatment with hMG was commenced, and treatment with this dose of buserelin continued until the day of hCG administration. For induction of follicular growth, treatment was commenced with three ampoules of hMG daily for at least 6 days. Further dosages were individualized on the basis of ultrasound examinations and oestradiol levels, until an adequate ovarian response had been obtained.

Human chorionic gonadotrophin, 5000 IU, was administered when three follicles > 14 mm in diameter were detected on ultrasound scan of the ovaries, with at least one ≥ 17 mm in diameter, in the presence of serum oestradiol concentration greater than 1500 pmol/l. Oocyte recovery was performed 35 h later, using transvaginal ultrasound-directed follicle aspiration. The technique of IVF, culture of oocytes and embryos, fertilization, and ET were as described by Owen and co-workers[8].

Follicular fluid for analysis of IGF–1 was collected by carefully selecting the clear portion of the aspirate using three separate tubes. The fluid was subsequently centrifuged, separated from cellular particles, and the supernatant stored frozen at −20°C. Only follicular fluid obtained from follicles containing an oocyte were used for measurement of IGF–1 concentrations. In total, 44 follicular fluid samples were analyzed: 26 from GH–augmented cycles and 18 from placebo cycles.

During the period of this study, up to four embryos that had shown evidence of normal cleavage were transferred to the uterus after 48 h of culture. Remaining embryos of sufficient quality were cryopreserved. Luteal-phase support was given to all patients in the form of hCG 5000 IU on the day of embryo transfer and again 3 days later.

The mean (± SD) age of the patients was 32.4 ± 3.0 years in the GH group, which was not significantly different from that in the placebo group (33.5 ± 2.6 years). The mean body mass index ($=$ weight (kg)/ height (m)^2) was also not significantly different between the two groups. The various clinical diagnoses were equally distributed: 11 patients had tubal damage, seven had unexplained infertility, three had oligospermia, two had male antibodies and two failed donor insemination.

Of the 25 patients, 18 were diagnosed as having PCOs, based on ovarian morphology. Pretreatment serum LH concentrations were not significantly different between those with ultrasound–diagnosed PCOs (median of 6.2 IU/l, range 3.1–25.8) and those with normal ovaries (median of 5.5 IU/l, range 3.7–6.8).

RESULTS

Comparing the results of the prestudy cycles in patients who went on to receive placebo or GH (Table 2), we found no significant difference between the groups in the dose of hMG used, number of days of treatment or in the number of follicles which developed. There were also no significant differences in the number of oocytes which were fertilized, or in the number of embryos which cleaved. There were, however, more oocytes collected in the prestudy cycles of the patients who went on to receive GH (median 6, range 0–8) than in those who went on to receive placebo (median 3.5, range 2–6), $p < 0.03$.

Comparing the results between the two treatment groups (Table 3), in those women who were randomized to receive GH, the total dose of

Table 2 Group comparison of the prestudy cycles of patients who subsequently received growth hormone (GH) or placebo. Values are medians with ranges in parentheses; hMG, human menopausal gonadotrophin; hCG, human chorionic gonadotrophin

| | *Prestudy cycles* | | |
	Placebo group (*n* = 12)	*GH group* (*n* = 13)	*p*
Days of hMG treatment	8.0 (6–12)	8.0 (7–10)	0.81
hMG/cycle (ampoules)	33.5 (18–64)	28.0 (14–60)	0.15
Follicles ≥ 14 mm (day of hCG)	4.0 (2–6)	4.0 (2–7)	0.29
Oocyte collected	3.5 (2–6)	6.0 (0–8)	0.03
Oocyte fertilized	1.5 (0–3)	2.0 (0–4)	0.52
Embryos cleaved	1.0 (0–3)	1.0 (0–3)	0.40
Embryos replaced	1.0 (0–3)	1.0 (0–3)	0.76

Table 3 Results of ovarian stimulation comparing growth hormone (GH) and placebo cycles. Values are medians with ranges in parentheses; hMG, human menopausal gonadotrophin; hCG, human chorionic gonadotrophin

	With placebo (*n* = 12)	*With GH* (*n* = 13)	*p*
Days of hMG treatment	8.5 (6–12)	8.0 (6–13)	0.73
hMG/cycle (ampoules)	36.0 (24–100)	28.0 (20–85)	0.05
Follicles ≥ 14 mm (day of hCG)	5.0 (2–9)	8.0 (3–19)	0.15
Oocyte collected	5.0 (2–13)	11.0 (2–16)	0.11
Serum oestradiol (pmol/l)	3139 (1495–8440)	3804 (1397–8960)	0.27
Oocyte fertilized	3.0 (0–6)	5.0 (0–11)	0.04
Oocyte cleaved	2.0 (0–5)	4.0 (0–10)	0.06
Embryos replaced	2.5 (0–4)	4.0 (0–4)	
Number of pregnancies	1	4	

hMG used was significantly less (median 28 ampoules, range 20–85) than in the group receiving placebo (36 ampoules, range 24–100) ($p < 0.05$). No significant differences were found between the two groups when

Table 4 Group comparison of placebo with growth hormone- (GH-) treated cycles in patients with polycystic ovary syndrome. Values are medians with ranges in parentheses; hMG, human menopausal gonadotrophin; hCG, human chorionic gonadotrophin

	Placebo (*n* = 8)	*GH* (*n* = 10)	*p*
hMG/cycle (ampoules)	36.0 (24–100)	27.5 (20–42)	0.01
Follicles ≥ 14 mm (day of hCG)	6.5 (4–9)	8.5 (4–19)	0.05
Serum oestradiol (pmol/l)	3991 (1613–8440)	5810 (2355–8960)	0.17
Oocyte collected	7.0 (2–13)	11.5 (6–16)	0.03
Oocyte fertilized	3.5 (0–6)	6.0 (5–11)	0.004
Oocyte cleaved	3.0 (0–5)	5.5 (3–10)	0.02
Embryos replaced	3.5 (0–4)	4.0 (3–4)	
Number of pregnancies	1	4	

comparing the number of follicles ≥ 14 mm in diameter on the day of hCG administration, and in the number of oocytes collected, although more oocytes were fertilized from the patients receiving GH than from those receiving placebo ($p < 0.04$).

When the subgroup of 18 patients with ultrasound-diagnosed PCOs was analyzed (Table 4), a similar pattern was observed. Group analysis of the eight PCO patients receiving placebo and the ten receiving GH showed that, despite using a significantly smaller dose of hMG ($p < 0.01$) in the patients receiving GH, more follicles developed ($p < 0.05$) and more oocytes were collected ($p < 0.03$), more of which were fertilized ($p < 0.004$) and cleaved ($p < 0.02$). There was, however, no significant difference in the maximum serum oestradiol concentrations between the treatment groups. Comparing the number of follicles which developed in the prestudy cycles with the number that developed in the subsequent cycles involving the two different treatment modalities, a significant effect of placebo as well as GH was found. There was, however, a significantly greater improvement in those who received GH treatment ($p < 0.04$).

Serum IGF-1 concentrations rose significantly during the GH treatment cycles and peaked between the second and third injection, falling back into the normal range by the day of egg collection. There was no change in IGF-1 concentrations in the placebo group. Although there was no significant difference in mean serum IGF-1 concentrations on the day of oocyte collection (23.6 ± 7.0 nmol/l for the placebo group versus 28.9 ± 8.2 nmol/l for the GH group), follicular fluid IGF-1 concentrations were significantly higher in the patients treated with GH than in those treated with placebo (24.9 ± 6.9 versus 19.6 ± 3.3 nmol/l, respectively, $p < 0.04$). In both groups, follicular fluid IGF-1 concentrations were significantly lower, compared with serum concentrations (19.6 ± 3.3 versus 26.3 ± 7.0 nmol/l, respectively, for the placebo group, $n = 10$, $p < 0.0002$, and 24.9 ± 6.9 versus 28.9 ± 8.2 nmol/l, respectively, for the GH group, $n = 13$, $p < 0.01$).

Four women conceived after their treatment cycle with GH, and delivered two sets of twins and two singletons (a rate of 46% live births per aspiration). One patient had an ectopic pregnancy following a cycle with placebo treatment.

DISCUSSION

These results confirmed our earlier findings on the induction of ovulation and IVF-ET[9-11]. Co-treatment with GH increased the ovarian response to stimulation by gonadotrophin in patients who had previously been resistant to hMG. The results demonstrated a 22.3% reduction in the total dose of gonadotrophins required for ovarian stimulation, despite which the ovarian performance improved, as judged by the increase in the number of developing follicles and of oocytes retrieved.

When the results from the subgroup of patients with ultrasonically diagnosed PCOs were analyzed, we found an even greater reduction in the total dose of gonadotrophins required for ovarian stimulation, and an even greater increase in the number of follicles that developed. This result suggested that even though these patients had previously responded suboptimally to gonadotrophin stimulation, when co-treated with GH, they were potentially very sensitive to gonadotrophin therapy. The cause of the initial lack of an adequate response to treatment with LHRH analogues and gonadotrophins alone is unknown, but might be caused by inadequate intra-ovarian production of growth factors.

The data showed significantly higher concentrations of IGF-1 in the follicular fluid of patients treated with GH compared with placebo, although there was no significant difference in serum IGF-1 concentrations on the day of egg collection. These results differ from those of Volpe and colleagues[12] who found no significant difference in follicular fluid IGF-1 concentrations between GH-treated patients, compared with a control group. However, their study-group included only 4 patients. In our study, the presence of IGF-1 in the follicular fluid in both groups might suggest a role for this growth factor in the development of the follicle. The significant rise in the concentrations of IGF-1 in the follicular fluid following GH treatment would be consistent with a local production of IGF-1 by granulosa cells, as has been demonstrated in *in vitro* experiments[13]. However, it is known that, quantitatively, IGF-1 is synthesized mainly in the liver[14], and, in view of the significantly lower levels of IGF-1 in follicular fluid compared with serum found in our study and in that of Rabinovici and co-workers[15], an extra-ovarian source may be the main contributor to follicular fluid IGF-1.

These and other studies[9–12,16–19] therefore show that co-treatment with GH augments the gonadal response to stimulation by gonadotrophins. The results of the studies that demonstrated a reduction in the total dose of gonadotrophins required for inducing ovulation have obvious implications for programmes of fertility treatment that require ovarian stimulation. They do, however, raise two important questions: whether the effect is physiological and represents a form of replacement treatment or whether it is a pharmacological effect. In the first studies we used very large doses of GH (about six times the dose used in the replacement treatment for short stature) but our subsequent experience suggests that the effect can be obtained with much smaller amounts of GH. Thus, the study described by Homburg and colleagues[16] showed that a single injection of 24 IU GH on the first day of gonadotrophin therapy was also effective in increasing ovarian sensitivity. A comparison of the results of a single-dose of co-treatment with GH and with the 6-dose GH protocol demonstrated an intermediate, albeit significant, reduction in the required number of ampoules, duration of treatment and daily effective dose of hMG. This result strongly suggests a dose-dependent response.

The optimal dose of GH treatment needed to increase ovarian sensitivity to gonadotrophin therapy will soon emerge from the results of a multicentre dose ranging study organized by Novo-Nordisk. We think that the results of our own studies probably represent the medical aspect of the essential physiological connection between growth and development.

Clearly, it is essential that the processes of growth and development are intimately connected so that the organism does not attempt to reproduce when it is physically too small for the mechanics of reproduction or birth to take place. Studies from our own group have shown that the application of pulsatile LHRH to children with delayed puberty causes an increase in the rate of growth as one of the earliest phenotypic expressions of treatment[20,21]. In our opinion, therefore, the studies described above represent the clinical application of these physiological observations.

The second question raised by these observations is whether the effect of GH we observed is exerted directly on the gonad or mediated through the IGFs. The knowledge that IGF-1 can modulate both basal and gonadotrophin-induced secretion of steroids in granulosa cells[4] suggests that IGF-1 may play an important role in the development and maturation of the ovarian follicle. The oocyte collection procedure provides access to the intra-ovarian environment, albeit in super-ovulation cycles, allowing us to determine local concentrations of these factors within the ovary, and to evaluate the possibility of ovarian secretion.

Human follicular fluid contains IGF-1, with significantly higher concentrations in dominant than in cohort follicles[22]. Based on the above-mentioned data and the notion that IGF-1, like GH[23] is capable of augmenting the action of FSH in the human ovary, we hypothesize that GH-FSH synergy may be due in part to the ability of GH to increase IGF-1 production and augmentation of FSH action. Whereas the precise therapeutic role of GH in induction of ovulation has yet to be defined, further studies to select the most appropriate group of patients to be treated, to explore the mechanism of action of GH, to define the role of IGF-1 in the process of human follicular development and to determine the minimum dose of GH that sensitizes the human ovary are, of course, required. We think, however, that the reality of GH interaction with gonadotrophins is now established.

Finally there is the practical consideration of which patients should be considered for co-treatment with GH. As regards those with anovulatory infertility, it is those with gonadotrophin and oestrogen deficiency who seem to benefit most. What is certain is that women with incipient or manifest primary ovarian failure do not respond[24]. Of the patients undertaking assisted fertility, it seems that a relatively select group of women benefit – that is, women with polycystic ovaries who are undergoing ovarian stimulation using an LHRH analogue and who prove resistant to stimulation by gonadotrophin. Again, it is crucial to emphasize that co-treatment with GH of women with ovarian failure is doomed to failure. On the other hand, judicious use of this hormone may provide the clinician with an additional form of treatment for the otherwise resistant case.

ACKNOWLEDGEMENTS

It is a pleasure to acknowledge the continued support of Dr Hanne Ostergaard of Novo-Nordisk, to whom we express our thanks for generous supplies of Norditropin. The studies of the use of growth hormone in the context of IVF/ET were only possible through the offices of Dr Bridgett Mason of The Hallam Medical Centre. I thank Dr Elizabeth Owen, Dr Roy Homburg and Dr Zeev Shoham for their clinical contributions. None of these studies would have been possible without the continued help of Ms Anita Patel and her colleagues in the Department of Ultrasonography at The Middlesex Hospital.

REFERENCES

1. Advis, J.P.S., White, S. and Ojeda, S.R. (1981). Activation of growth hormone short loop negative feedback delays puberty in the female. *Endocrinology*, **108**, 1343–52
2. Laron, Z., Sarel, R. and Pertzelan, A. (1980). Puberty in Laron type dwarfism. *Eur. J. Pediatr.*, **134**, 79–83
3. Shiekholislam, B.M. and Stempfel, R.S. (1972). Hereditary isolated somatotropin deficiency: effects of human growth hormone administration. *Pediatrics*, **49**, 362–74
4. Adashi, E.Y., Resnick, C.E. and D'Ercole, A.J. (1985). Insulin-like growth factors as intraovarian regulators of granulosa cell growth and function. *Endocr. Rev.*, **6**, 400–20
5. Hernandez, E.R., Roberts, C.T., LeRoith, D. *et al.* (1989). Rat ovarian insulin-like growth factor I (IGF-I) gene expression in granulosa cell-selective: 5' untranslated mRNA variant representation and hormonal regulation. *Endocrinology*, **125**, 572–4
6. Oliver, J.E., Aitman, T.J., Powell, J.F. *et al.* (1989). Insulin-like growth factor I gene expression in the rat ovary is confined to the granulosa cells of developing follicle. *Endocrinology*, **124**, 2671–8

7. Adams, J., Franks, S., Polson, D.W. *et al.* (1985). Multifollicular ovaries: clinical and endocrine features and response to pulsatile gonadotrophin releasing hormone. *Lancet*, **2**, 1375–8

8. Owen, E.J., Davies, M.C., Kingsland, C.R. *et al.* (1989). The use of a short regimen of buserelin, a gonadotrophin-releasing hormone agonist, and human menopausal gonadotrophin in assisted conception cycles. *Hum. Reprod.*, **4**, 749–54

9. Homburg, R., Eshel, A., Abdalla, H.I. *et al.* (1988). Growth hormone facilitates ovulation induction by gonadotrophins. *Clin. Endocrinol.*, **29**, 113–17

10. Homburg, R., West, C., Torresani, T. *et al.* (1990). Co-treatment with human growth hormone and gonadotrophins for induction of ovulation: a controlled clinical trial. *Fertil. Steril.*, **53**, 254–60

11. Owen, E.J., West, C. and Mason, B.A. (1991). Co-treatment with growth hormone of sub-optimal responders in IVF-ET. *Hum. Reprod.*, **6**, 524–8

12. Volpe, A., Coukos, G., Barreca, A. *et al.* (1989). Ovarian response to combined growth hormone-gonadotrophin treatment in patients resistant to induction of superovulation. *Gynecol. Endocrinol.*, **3**, 125–34

13. Steinkampf, M.P., Mendelson, C.R. and Simpson, E.R. (1988). Effects of epidermal growth factor and insulin-like growth factor-1 on the levels of mRNA encoding aromatase cytochrome P-450 of human ovarian granulosa cells. *Mol. Cell. Endocrinol.*, **59**, 93–7

14. D'Ercole, A.J., Stiles, A.D. and Underwood, L.E. (1984). Tissue concentrations of somatomedin C: further evidence for multiple sites of synthesis and paracrine or autocrine mechanisms of action. *Proc. Natl. Acad. Sci. USA*, **81**, 935–9

15. Rabinovici, J., Dandekar, P., Angle, M.J. *et al.* (1990). Insulin-like growth factor I (IGF-I) levels in follicular fluid from human preovulatory follicles: correlation with serum IGF-I levels. *Fertil. Steril.*, **54**, 428–33

16. Homburg, R., West, C., Torresani, T. *et al.* (1990). A comparative study of single-dose growth hormone therapy as an adjunct to gonadotrophin treatment for ovulation induction. *Clin. Endocrinol.*, **32**, 781–5

17. Owen, E.J., Shoham, Z., Mason, B.A. *et al.* (1991). Cotreatment with growth hormone, following pituitary suppression, for ovarian stimulation in *in-vitro* fertilization: a randomized, double-blind, placebo-control trial. *Fertil. Steril.*, **56**, 1104–10

18. Blumenfeld, Z. and Lunenfeld, B. (1989). The potentiating effect of growth hormone on follicle stimulation with human menopausal gonadotrophins in a panhypopituitary patient. *Fertil. Steril.*, **52**, 328–31

19. Ibrahim, Z.H.Z., Matson, P.L., Buck, P. *et al.* (1991). The use of biosynthetic human growth hormone to augment ovulation induction with

buserelin acetate/human menopausal gonadotrophin in women with a poor ovarian response. *Fertil. Steril.*, **55**, 202–4

20. Stanhope, R., Brook, C.G.D., Pringle, P.J., Adams, J. and Jacobs, H.S. (1987). Induction of puberty by pulsatile gonadotrophin releasing hormone. *Lancet*, **2**, 552–5

21. Darendeliler, F., Hindmarsh, P.C., Preece, M.A. *et al.* (1990). Growth hormone increase rate of pubertal maturation. *Acta Endocrinol.*, **122/3**, 414–16

22. Eden, A.J., Jones, J., Carter, G.D. *et al.* (1989). A comparison of follicular fluid levels of insulin-like growth factor-I in normal dominant and cohort follicles, polycystic and multicystic ovaries. *Clin. Endocrinol.*, **29**, 327–9

23. Jia, X.C., Kalmijn, J. and Hsueh, A.J.W. (1986). Growth hormone enhances follicle-stimulating hormone-induced differentiation of cultured rat granulosa cells. *Endocrinology*, **118**, 1401–9

24. Homburg, R., West, C., Ostergaard, H. and Jacobs, H.S. (1991). Combined growth hormone and gonadotrophin treatment for ovulation induction in patients with non responsive ovaries. *Gynecol. Endocrinol.*, **5**, 33–6

10

Ovarian surgery

A. Abdel Gadir

INTRODUCTION

Goldzieher and Green[1] described the polycystic ovarian syndrome (PCOS) as a non-tumourous dysfunctional condition of the ovaries characterized by luteinizing hormone (LH) dependent hypersecretion of androgens from hyperplastic theca and stroma cells. The histological picture seen in an individual case of PCOS depends on the relative number of the follicular and atretic cysts, the degree of stromal hyperplasia and the number of primordial follicles present. At one end of the spectrum, ovaries contain more follicular cysts, with few atretic ones, minimal stromal hyperplasia and abundant primordial follicles. This group may be represented ultrasonically by the multifollicular ovarian pattern. At the other extreme, there are numerous atretic follicles with a small number of follicular cysts and marked stromal hyperplasia, with islands of luteinized stromal cells giving the typical ultrasonic image of PCOS. Accordingly, a single histological or sonographic picture may not represent the whole spectrum seen in women with PCOS. Furthermore, the mode of clinical presentation[2] and the response to the different treatment modalities may reflect the prevalent histological picture in the individual patient. This last point is evidenced by the fact that patients with larger ovaries, and presumably more antral and pre-antral follicles for recruitment, had a better clinical response to induction of ovulation with human menopausal gonadotrophins[3].

111

Biochemically, PCOS is characterized by one or more of the following criteria: high luteinizing hormone (LH) basal values or a high LH : follicle stimulating hormone (FSH) ratio, elevated androgens, deficient acyclic oestradiol production with disturbed oestradiol : oestrone ratio. Hormone pulse studies showed that women with PCOS have high LH and testosterone pulse-amplitude values, with or without a high pulse frequency[2,4]. Nevertheless, we showed that there was no universal LH pulse pattern common to all women with PCOS[5]. Some patients had high, and others low LH pulse pattern components and few had inverted LH : FSH ratio, despite our stringent criteria for patient recruitment. This reflected a non-persistent or variable endocrine milieu at different times in the same women, despite synchronized blood sampling relative to the menstrual cycle. Moreover, using Pearson's product-moment statistics, we could not find a significant correlation between LH and testosterone values or ovarian volume in patients with PCOS[6]. This discrepancy may be due to the fact that once PCOS develops, the ovary assumes a primary role in androgen hypersecretion[1,7]. This may reflect the increased sensitivity of theca cells to LH, as shown by increased production of androstenedione after LH stimulation in patients with PCOS, in comparison to normal women[8]. It may, in addition, be a reflection of the build-up of the secondary interstitial tissue in the ovarian stroma following the progressive follicular atresia. These points may explain the favourable endocrine and clinical changes that followed local manipulation of the ovaries during the different surgical modalities used in the treatment of patients with PCOS.

HISTORICAL CONSIDERATIONS

Various surgical procedures have previously been used for the treatment of patients with PCOS. Ovarian wedge resection[9], extroversion of the ovaries[10], decortication[11] and medullectomy[12] have been described. Resumption of menstrual function following unilateral oophorectomy without manipulation of the other ovary was reported, but this has not been confirmed in three subsequent cases[13]. However, ovarian wedge resection remained the most popular surgical treatment for patients with PCOS until clomid became available[14] when surgical treatment was reserved for clomid non-responders. This may be a reflection of the fact that the initial reports documented by Stein and Cohen[15] were not confirmed by other investigators[16–18], and were due to the high recurrence rate following wedge resection[19]. These points were compounded by a high incidence of peritoneal adhesions[18,20] and a reported eightfold increase in the incidence of ectopic pregnancy[21] following surgery. Furthermore, the introduction of human menopausal gonadotrophins for the treatment of patients with anovulatory infertility reduced the need for ovarian wedge resection when treating patients with PCOS resistant to clomid[22].

CURRENT PRACTICE

Anti-oestrogens are used as the primary medication for treating anovulatory patients with PCOS. However, 15% of such patients treated with clomiphene citrate remain anovulatory[23] and will need alternative treatment modalities. This may be in the form of injectable gonadotrophins which need serial endocrine and ultrasonic monitoring with the inherent risks of hyperstimulation and multiple pregnancies. Accordingly, different laparoscopic surgical methods have been described recently for the treatment of patients with PCOS. Ovarian electrocautery[24], multiple ovarian punch biopsies[25] and laser vaporization of the ovarian capsule and atretic follicles[26–28] have been described. These modalities were introduced to circumvent the need for ovarian wedge resection and the inevitable adhesion formation that follows this procedure with ensuing tubal infertility. All reports published to date have concentrated on the treatment of anovulatory patients who were not responsive to clomiphene citrate therapy. Different rates of successful induction of ovulation and pregnancy were reported by different investigators, reflecting the different criteria used for diagnosing PCOS and the different methods used for reporting results. Moreover, there are no reports yet available which described the efficacy of these procedures in treating hirsutism, obesity or recurrent miscarriages.

This review will concentrate on ovarian electrocautery, since other methods of laparoscopic ovarian surgery are not yet practised widely. Several questions need to be addressed to assess the efficacy and safety of the procedure, as follows:

(1)   How effective and safe is the procedure?

(2)   Who should we treat?

(3)   How can we maximize the results?

(4)   What is its mode of action?

(5)   Prospective thinking!

HOW EFFECTIVE AND SAFE IS THE PROCEDURE?

Various investigators reported a decrease in the level of LH and testosterone following ovarian electrocautery[29–32]. Figure 1 shows the 6-h mean values of LH, FSH, testosterone and oestradiol following ovarian electrocautery or eight weeks of intranasal buserelin (Hoechst, Germany) in a dose of 800 µg/day. There was an equivalent diminution in the level of LH and testosterone following both procedures, with an increase in the level of FSH after ovarian electrocautery[33]. Moreover, we compared the effect of ovarian electrocautery versus buserelin medication in the response of patients with PCOS to human menopausal gonadotrophin (hMG) therapy[34]. There was no difference in the ovulation or pregnancy rate between the two groups. However, the number of cycles with multiple dominant follicles, and accordingly the risk of hyperstimulation and multiple pregnancy, was lower in the ovarian electrocautery group. Furthermore, the luteal phase serum testosterone and miscarriage rate were also lower in the group pretreated with ovarian electrocautery. A quantitative assessment of the clinical response against gonadotrophin therapy showed unsupplemented ovarian electrocautery to be equally effective to hMG and pure FSH in treating patients with PCOS[3]. Successful ovulation was induced in 71.4% of the cycles after electrocautery, 70.6% after hMG and 66.7% after pure FSH medication. The pregnancy rate per ovulatory cycle was 13.3%, 17.9% and 13.2%, respectively ($\chi^2 = 0.883$, $p > 0.5$) and the 6-cycle cumulative pregnancy rate was 52.1%, 55.4% and 38.3%, respectively. All pregnancies following ovarian electrocautery were singleton whereas 20% of pregnancies following gonadotrophin therapy were multiple. The miscarriage rate in the three groups was 21.4%, 53.3% and 40% respectively ($\chi^2 = 3.128$, $p = 0.2093$). These results showed that ovarian electrocautery is as effective as gonadotrophin therapy in the treatment of patients with PCOS and has the further advantage of a lower risk of multiple pregnancies and hyperstimulation. These advantages are further enhanced by the low risk

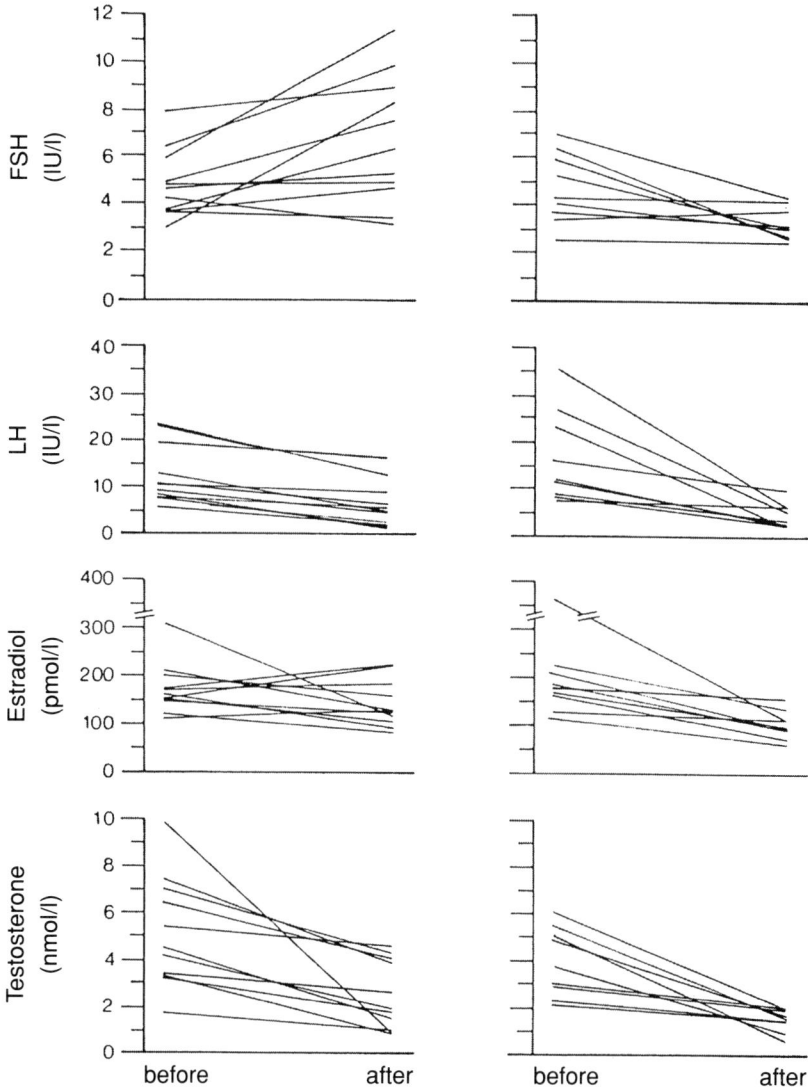

Figure 1 Mean hormone levels (6-h means) before and after ovarian electrocautery (left) or 8 weeks of intranasal buserelin medication (right) in patients with polycystic ovarian syndrome. FSH, follicle stimulating hormone; LH, luteinizing hormone

of adhesion formation reported after ovarian electrocautery[35], and laser vaporization[28]. Nevertheless, we advocate the use of a sharp needle for electrocautery and the application of the needle at a right angle to prevent slit cauterization and to reduce the damaged area on the surface of the ovary. This procedure should be supplemented with the creation of artificial ascites with dextran to keep tissues afloat, in the immediate postoperative period. All these points are in agreement with the current microsurgical concepts used to reduce the formation of adhesions.

WHO SHOULD WE TREAT?

We showed that 24% of patients with PCOS resistant to clomid therapy did not respond to unsupplemented ovarian electrocautery[3]. Table 1 shows the endocrine and biophysical attributes of responders compared to non-responders. Clinical responders had a high basal LH level (> 12 IU/l), but body mass index, ovarian volume and pretreatment testosterone level had no prognostic value. This point is further supported by the fact that the 6-h mean LH value, following 10-min blood sampling, was higher in responders than in non-responders. Furthermore, clinical responders showed a greater reduction in the level of LH after surgery than non-responders (Figure 2).

These findings stress the importance of LH in the genesis or maintenance of PCOS. This is further supported by the observation that LH level was rising in successive cycles in women who became refractory during the follow-up period (Figure 3). Accordingly, ovarian electrocautery should be offered to patients with PCOS and high LH levels (> 12 IU/l). Furthermore, patients who hyper-responded to gonadotrophin therapy in previous cycles may benefit from ovarian electrocautery, especially since the number of multiple dominant follicles at midcycle was shown to be lower after ovarian electrocautery plus hMG than with buserelin plus hMG[34]. Similarly, since the miscarriage rate was shown to be lower following supplemented or unsupplemented electrocautery versus gonadotrophin therapy with or without pituitary desensitization, one may offer ovarian cautery to those patients who have had a history of previous miscarriages. However, more work needs to be done to verify this point.

Table 1 Basal biophysical and endocrine attributes of responders and non-responders before ovarian electrocautery. Figures are means with 95% confidence intervals in parentheses; FSH, follicle stimulating hormone; LH, luteinizing hormone

	Responders (n = 22)		Non-responders (n = 7)	
Age (years)	27.05	(25.49–28.60)	29.43	(26.51–32.35)
Body mass index (kg/m²)	29.42	(27.85–30.99)	27.30	(23.80–30.80)
Ovarian volume (cm³)	16.87	(14.71–19.04)	15.55	(10.76–20.35)
FSH (IU/l)	5.90	(5.25–6.55)	5.45	(3.36–7.55)
LH (IU/l)	16.42	(12.88–19.97)★	9.77	(6.58–12.96)
Testosterone (nmol/l)	3.71	(3.16–4.26)	4.54	(3.36–5.72)
Oestradiol (pmol/l)	167.90	(137.86–197.95)	170.71	(112.99–228.44)

★ Significant difference from non-responders ($p < 0.05$)

HOW CAN WE MAXIMIZE THE RESULTS?

Figure 4 shows the endocrine profile in responders versus non-responders during the 1st week after surgery. It is evident that the two groups differed only in their FSH response. Follicle stimulating hormone levels increased significantly in the first day after surgery whereas luteinizing hormone values showed a continuous decline which became significantly different on the fifth day following electrocautery in responders only. Both responders and non-responders had a significant and equivalent decline in testosterone level starting on the first postoperative day. Oestradiol levels showed no significant change within or between the two groups. This finding can be used to maximize the effect of the procedure by monitoring all patients following surgery for 1 week. Those patients who do not show an increase in FSH level with a reversal in the LH:FSH ratio should be supplemented with clomid, which has an improved effect following cautery[3,32,36]. Furthermore, ultrasonic or urinary LH monitoring should be performed following the start of the cervical mucus biological shift, to monitor ovulation and to assist with timed sexual intercourse and probably increase the pregnancy rate. Thirdly, serial measurement of LH level on the 5th day of the subsequent cycles may detect a rising level, which is an early sign for the recurrence of

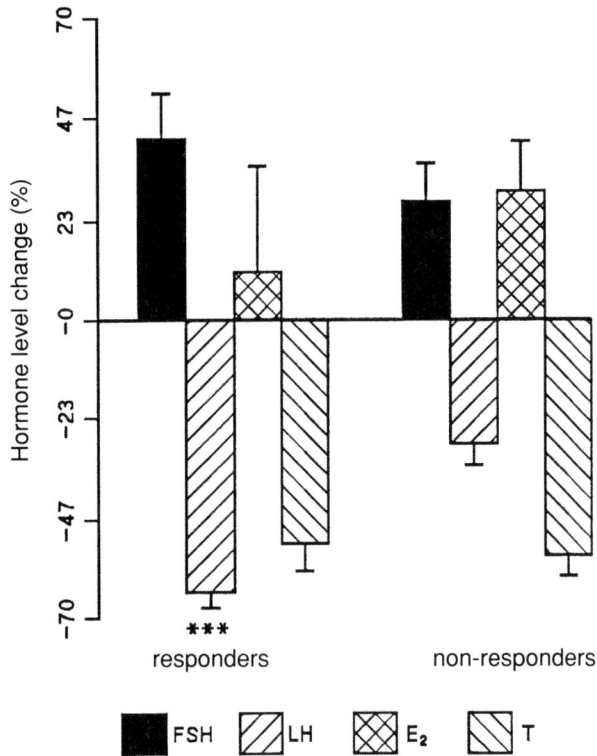

Figure 2 Changes in the 6-h mean hormone level values (%) before and after ovarian electrocautery in responders and non-responders. Error bars represent SEMs; FSH, follicle stimulating hormone; LH, luteinizing hormone; E_2, oestradiol; T, testosterone; ***, $p = 0.001$

anovulation (Figure 3) and, accordingly, an indication for clomid supplementation. All these points may increase the pregnancy rate since there is an improvement in the endocrine profile following ovarian electrocautery, even in clinical non-responders[37]. Alternatively, ovarian electrocautery may be repeated in those patients who become refractory after an initially favourable response; this is especially so since the procedure has been shown to be safe and uncomplicated by any significant adhesion formation.

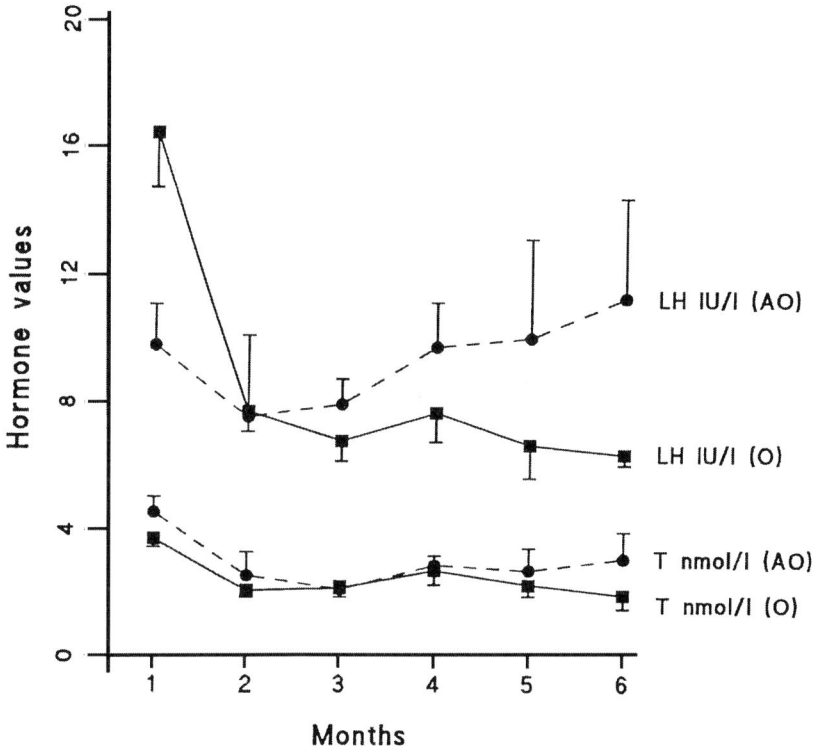

Figure 3 Basal hormone levels on the 5th day during ovulatory (O) and anovulatory (AO) cycles. Error bars respresent SEMs; LH, luteinizing hormone; T, testosterone

WHAT IS THE MODE OF ACTION?

Like ovarian wedge resection, the exact mode of action of laparoscopic ovarian surgery is not yet settled. The improved response to clomid therapy, with its central effect, and to hMG medication, with its local ovarian action, suggests that both central and ovarian mechanisms are involved. Most investigators have reported a reduction in LH pulse amplitude, which is suggestive of a modified pituitary response to LHRH stimulation[25,37]. However, Ohkouchi and colleagues[38] reported a reduced LH pulse frequency, following ovarian wedge resection, which is suggestive of an alteration in the hypothalamic electrical activity necessary to

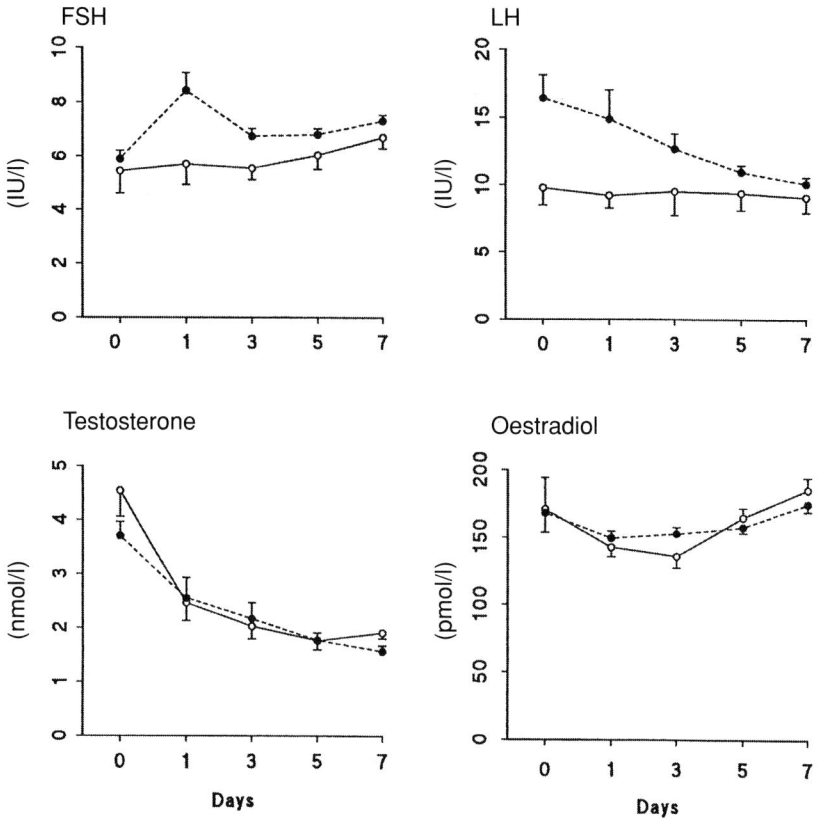

Figure 4 Endocrine profile in responders (●—●) versus non-responders (o—o) during the 1st week after surgery. Error bars represent SEMs; FSH, follicle stimulating hormone; LH, luteinizing hormone

generate LHRH pulses. This central effect may be due to the reduction in androgen concentration without oestrogen intervention, especially because the early decline in the level of LH following surgery was not matched by a significant decrease in the concentration of oestradiol[37] or oestrone[39]. This central effect is coupled by a reduction in the intra-ovarian androgen concentration which facilitated the response of patients with PCOS to hMG medication following ovarian electrocautery. This is shown by an attenuated testosterone response after human chorionic gonadotrophin (hCG) administration in these patients[34]. Furthermore,

Sakata and colleagues[39] reported a decrease in the concentration of the bioactive LH component following ovarian electrocautery. The issue of mode of action of ovarian surgery in the treatment of patients with PCOS was reviewed by Vaughan Williams[40]. She showed that the common factor to ovarian wedge resection, laser vaporization and electrocautery was the drainage of the androgen and inhibin-rich contents of the atretic ovarian cysts, rather than the removal of part of the interstitial tissue or breaching the integrity of the ovarian tunica.

PROSPECTIVE THINKING

It is evident that laparoscopic ovarian surgery has proved to be a useful method for treating patients with PCOS. Currently, only patients with anovulation resistant to clomid therapy are offered this type of treatment. However, with the sustained effect of ovarian cautery on the level of serum androgens, the potential for this procedure in treating patients with clinical hyperandrogenism should be further investigated. Obese patients with PCOS may find it difficult to lose weight because of the anabolic effect of high androgen levels. This cycle can be broken, as obese patients were shown to respond favourably to ovarian electrocautery[3]. This method may further help with the treatment of dysfunctional uterine bleeding and avoid the need for repeated courses of hormonal therapy and numerous unnecessary applications of cervical dilatation and uterine curettage. Currently, we are investigating the potential of ovarian electrocautery in assisted reproduction.

CONCLUSION

With the current surge in use of pelvic ultrasonography as the sole criterion for the diagnosis of polycystic ovaries and the wide use of laparoscopic ovarian surgery for the treatment of patients with PCOS, more conservative criteria should be used in both areas to prevent overtreatment. We have shown that women with PCOS and high LH levels showed a better response than patients with ultrasonically visualized PCOS and normal LH levels but with a high LH:FSH ratio. Closer monitoring, without over-indulgence, may improve the pregnancy rate as it may help with timed intercourse. It may also help in the early detection of patients in need of clomid supplements, thus converting anovulatory cycles into fertile ones.

REFERENCES

1. Goldzieher, J.W. and Green, J.A. (1962). The polycystic ovary: clinical and histological features. *J. Clin. Endocrinol. Metab.*, **22**, 325–38

2. Yen, S.S.C. (1986). Chronic anovulation caused by peripheral endocrine disorders. In Yen, S. S. C. and Jafe, W. B. (eds.) *Reproductive Endocrinology, Physiology, Pathophysiology and Clinical Management*, pp. 441–9 (Philadelphia: W.B. Saunders)

3. Abdel Gadir, A., Mowafi, R.S., Alnaser, H.M.I., Alrashid, A.H., Alonezi, O.M. and Shaw, R.W. (1990). Ovarian electrocautery versus human menopausal gonadotrophins and pure follicle stimulating hormone therapy in the treatment of patients with polycystic ovarian disease. *Clin. Endocrinol.*, **33**, 585–92

4. Waldstreicher, J., Santoro, N.F., Hall, J.E., Filicori, M. and Crowly, W.F. (1988). Hyperfunction of the hypothalamo-pituitary axis in women with polycystic ovarian disease: Indirect evidence for partial gonadotrophin desensitization. *J. Clin. Endocrinol. Metab.*, **66**, 165–72

5. Abdel Gadir, A., Khatim, M.S., Mowafi, R.S., Alnaser, H.M.I., Alzaid H.G.N. and Shaw, R.W. (1991). Polycystic ovaries: do these represent a specific endocrinopathy? *Br. J. Obstet. Gynaecol.*, **98**, 300–305

6. Abdel Gadir, A., Khatim, M., Mowafi, R., Alnaser, H.M.I., Muharib, N.S. and Shaw, R.W. (1992). Implications of ultrasonically diagnosed polycystic ovaries (1) Correlations with basal hormonal profiles. *Hum. Reprod.*, **7**, 912–14

7. Fisher, E.R., Gregorio, R., Stephan, T., Nolan, S. and Donowski, T.S. (1974). Ovarian changes in women with morbid obesity. *Obstet. Gynecol.*, **44**, 839–44

8. McNatty, K.P., Moor-Smith, D., Makris, A., De Grazia, C., Tulchinsky, D., Osathanondh, R., Schiff, I. and Ryan, K.J. (1980). The intraovarian sites of androgens and oestrogen formation in women with normal and hyperandrogenic ovaries as judged by *in vitro* experiments. *J. Clin. Endocrinol. Metab.*, **50**, 755–63

9. Stein, I.F. and Leventhal, M.L. (1935). Amenorrhoea associated with bilateral polycystic ovaries. *Am. J .Obstet. Gynecol.*, **29**, 181–91

10. Bailey, K.V. (1937). The operation of extroversion of the ovaries for functional amenorrhoea especially of the secondary type. *J. Obstet. Gynaecol. Br. Emp.*, **44**, 637–49

11. Reycraft, J.L. (1938). Surgical treatment of ovarian dysfunctions. *Am. J. Obstet. Gynecol.*, **35**, 505–12

12. Allen, W.M. and Woolf, R.B. (1959). Medullary resection of the ovaries in the Stein Leventhal syndrome. *Am. J. Obstet. Gynecol.*, **77**, 826–37

13. Goldzieher, J.W. (1982). Polycystic ovarian disease. In Wallach, E.E. and Kempers, R.D. (eds.) *Modern Trends in Infertility and Conception Control*, Vol. 2, pp. 65–88 (Philadelphia: Harper and Row)

14. Greenblatt, R.B., Barfield, W.E., Jungck, E.C. and Ray, A.W. (1961). Induction of ovulation with MRL/41. *JAMA*, **178**, 101–4

15. Stein, I.F. and Cohen, M.R. (1939). Surgical treatment of bilateral polycystic ovaries, amenorrhoea and sterility. *Am. J. Obstet. Gynecol.*, **38**, 465–80

16. Goldzieher, J.W. and Axelrod, L.R. (1963). Clinical and biochemical features of polycystic ovarian disease. *Fertil. Steril.*, **14**, 631–53

17. Zarate, A., Henandez-Ayup, S. and Rios-Montiel, A. (1971). Treatment of anovulation in the Stein Leventhal Syndrome. Analysis of 90 cases. *Fertil. Steril.*, **22**, 188–93

18. Buttram, V.C. and Vaquero, C. (1975). Post-ovarian wedge resection adhesive disease. *Fertil. Steril.*, **26**, 874–6

19. Starup, J. (1976). Treatment of patients of polycystic ovarian syndrome. *Ugeskr. Laeger*, **138**, 2866–70

20. Kistner, R.W. (1969). Peri-tubal and peri-ovarian adhesions subsequent to wedge resection of the ovaries. *Fertil. Steril.*, **20**, 35–42

21. Adashi, E.Y., Rock, J.A., Guzick, D., Wentz, A.C., Jones, G.S., Jones, H.W. (1981). Fertility following bilateral ovarian wedge resection: A critical analysis of 90 consecutive cases of the polycystic ovary syndrome. *Fertil. Steril.*, **36**, 320–5

22. Lunenfeld, B. (1963). Treatment of anovulation by human gonadotrophins. *Int. J. Gynecol. Obstet.*, **1**, 153–67

23. Franks, S., Adams, J., Mason, H. and Polson, D. (1985). Ovulatory disorders in women with polycystic ovary syndrome. *Clin. Obstet. Gynaecol.*, **12**, 605–32

24. Gjönnaess, H. (1984). Polycystic ovarian disease treated by ovarian electrocautery through the laparoscope. *Fertil. Steril.*, **41**, 20–5

25. Sumioki, H., Utsunomyiya, T., Matsuoka, K., Korenaga, M. and Kadota, T. (1988). The effect of laparoscopic multiple punch resection of the ovary on hypothalamo–pituitary axis in polycystic ovary syndrome. *Fertil. Steril.*, **50**, 567–72

26. Huber, J., Hosmann, J. and Spona, J. (1988). Polycystic ovarian syndrome treated by laser through the laparoscope. *Lancet*, **2**, 215

27. Daniell, J.F. and Miller, W. (1989). Polycystic ovaries treated by laparoscopic laser vaporization. *Fertil. Steril.*, **51**, 232–6

28. Keckstein, G., Rossmanith, W., Spatzier, V., Kirstin, B. and Steiner, R. (1990). The effect of laparoscopic treatment of polycystic ovarian disease by CO_2 laser or Nd:YAG laser. *Surg. Endosc.*, **4**, 103–7

29. Aakvaag, A. and Gjönnaess, H. (1985). Hormonal response to electrocautery

123

of the ovary in patients with polycystic ovarian disease. *Br. J. Obstet. Gynaecol.*, **92**, 1258–64

30. Greenblatt, E. and Casper, R.F. (1987). Endocrine changes after laparoscopic ovarian cautery in polycystic ovarian syndrome. *Am. J. Obstet. Gynecol.*, **156**, 279–85

31. Gjönnaess, H. and Norman, N. (1987). Endocrine effects of ovarian electrocautery in patients with polycystic ovarian disease. *Br. J. Obstet. Gynaecol.*, **94**, 779–83

32. Armar, N., McGarrigle, H.H.G., Honour, J., Holownia, P., Jacobs, H.S. and Lachelin, G.C.L. (1990). Laparoscopic ovarian diathermy in the management of anovulatory infertility in women with polycystic ovaries: endocrine changes and clinical outcome . *Fertil. Steril.*, **53**, 45–9

33. Abdel Gadir, A., Khatim, M.S., Mowafi, R.S., Alnaser, H.M.I., Alzaid, H.G.N. and Shaw, R.W. (1990). Hormonal changes in patients with polycystic ovarian disease after ovarian electrocautery or pituitary desensitization. *Clin. Endocrinol.*, **32**, 749–54

34. Abdel Gadir, A., Alnaser, H.M.I., Mowafi, R. and Shaw, R.W. (1992). The response of patients with polycystic ovarian disease to human menopausal gonadotropin therapy after ovarian electrocautery or a luteinizing hormone-releasing hormone agonist. *Fertil. Steril.*, **57**, 309–13

35. Dabirashrafi, H., Mohamad, K., Behjatnia, Y. and Moghadami, N. (1991). Adhesions formation after ovarian electrocauterization on patients with polycystic ovarian syndrome. *Fertil. Steril.*, **55**, 1200–1

36. Van Der Weiden, R.M. and Alberda, A.T. (1987). Laparoscopic ovarian electrocautery in patients with polycystic ovarian disease resistant to clomiphene citrate. *Surg. Endosc.*, **1**, 217–19

37. Abdel Gadir, A., Khatim, M.S., Mowafi, R.S., Alnaser, H.M.I. and Shaw, R.W. (1991). Hormonal changes following laparoscopic ovarian electro-cautery in patients with polycystic ovarian syndrome. In Shaw, R.W. (ed.) *Advances in Reproductive Endocrinology*, Vol. 3, pp. 135–47 (Carnforth: Parthenon Publishing)

38. Ohkouchi, T., Tanaka, T., Oikawa, M., Sakuragi, N., Fujimoto, S. and Ichinoe, K. (1987). Effect of ovarian wedge resection and spironolactone administration on pulsatile LH release and steroid hormone levels in women with polycystic ovarian disease. *Acta Obstet. Gynaecol. Jpn.*, **39**, 626–32

39. Sakata, M., Tasaka, K., Kurachi, H., Terakawa, N., Miyake, A. and Tanizawa, O. (1990). Changes of bioactive luteinizing hormone after laparoscopic ovarian cautery in patients with polycystic ovarian syndrome. *Fertil. Steril.*, **53**, 610–13

40. Vaughan Williams, C.A. (1990). Ovarian electrocautery or hormone therapy in the treatment of polycystic ovary syndrome. *Clin. Endocrinol.*, **33**, 569–72

11

Premature luteinization

R. Fleming, M.E. Jamieson and J.R.T. Coutts

INTRODUCTION

Premature luteinization is the product of a positive feedback surge of luteinizing hormone (LH) on one or more immature follicles, leading to an increase in circulating progesterone concentrations above normal follicular phase values. Follicles may respond with increased progesterone output whether or not they have attained normal preovulatory dimensions (follicle diameter: 17–25 mm), but they may not be capable of normal ovulation. This phenomenon is a common response to the elevated circulating oestradiol concentrations encountered during stimulation of multiple follicular growth and was first observed by Gemzell in 1978[1].

In clinical terms, 'premature luteinization' is most frequently applied to LH surges occurring at inconvenient times in *in vitro* fertilization (IVF) programmes, the most important consequences of which include ovulation prior to oocyte retrieval and/or rapid rearrangements of operating times. However, there are additional important sequelae which have not been clearly defined.

It is the problems caused by premature luteinization in IVF cycles which have led to the widespread adoption of the techniques of follicular stimulation in conjunction with gonadotrophin releasing hormone analogues (GnRH-a) – to suppress LH fluctuations, minimize monitoring, and obviate short-term rearrangement of operative procedures.

125

ANIMAL STUDIES

The clearest demonstration of the complex nature of premature luteinization was the study in which does were injected with differing doses of human chorionic gonadotrophin (hCG) after follicular growth was induced by pregnant mare serum gonadotrophin[2]. At the lower hCG doses there was a differential effect on both follicles and oocytes, and oocyte nuclear maturation varied profoundly. The various oocyte abnormalities were dependent upon both the maturity of the individual follicle and the amount of hCG administered. Lower doses of hCG also induced failure of follicles to rupture properly, leading to luteinized unruptured follicles, enclosing oocytes at differing stages of nuclear maturation. This experiment highlighted the importance of both the correct timing of the luteinizing signal with respect to follicle maturity and the scale of the signal itself. It also indicated that a luteinizing signal was able to initiate follicle/luteal differentiation irrespective of follicle maturity and, in the process, it blocked subsequent follicle maturation leading to abnormalities of oocyte nuclear maturation.

HUMAN STUDIES

Unstimulated cycles

There is no evidence that the premature LH surge occurs in unstimulated cycles, although it is possible that it may constitute part of the aetiology of other abnormalities observed in patients with 'unexplained infertility'[3]. Figure 1 shows peripheral plasma oestradiol and progesterone profiles of an infertile patient with normal menstrual rhythm who demonstrated subnormal luteal phase progesterone concentrations following subnormal follicular phase oestradiol concentrations. It is conceivable that the profile of oestradiol was merely that of a normal follicle interrupted by a premature LH surge which caused the abrupt decline in oestradiol and initiated the subnormal surge of progesterone. However, subnormal pre-ovulation oestradiol profiles do not occur with a high frequency in normally cycling patients and, furthermore, they demonstrate a low recurrence rate in subsequent cycles[3]. The recurrence rate of deficient luteal phase progesterone profiles is significant, but

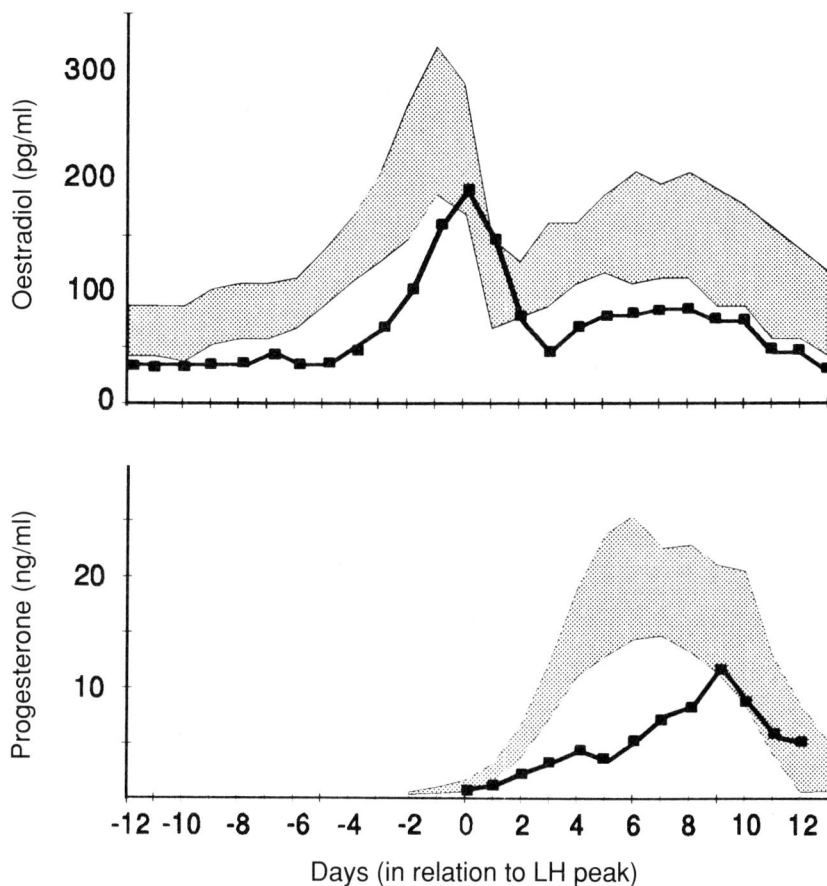

Figure 1 Hormone profiles of a patient displaying poor follicular maturation and deficient luteal phase oriented around the luteinizing hormone (LH) surge (day 0)

normal pre-ovulatory oestradiol and follicular diameter profiles were recorded in these cycles[3].

Luteinized unruptured follicles (identified in the rabbit experiments above) have been identified in normally cycling patients with infertility[4], although the incidence and possible causes are unclear. Their contribution to the aetiology of infertility remains debatable[5].

Stimulated cycles

During the induction of follicular development using exogenous gonadotrophins, negative feedback mechanisms are irrelevant and monofollicular
growth is rare. The corresponding elevated oestradiol profiles are capable
of eliciting a positive feedback LH surge at a time dictated by the
circulating oestradiol concentrations (with or without a gonadotrophin
surge attenuating factor) rather than by individual follicle maturity. During
the normal cycle, the LH surge is seen when the mean follicle diameter
(FD) is 20 mm and the oestradiol level is approximately 900 pmol/l (*ca.*
300 pg/ml)[6], but with induced multiple follicular development, the surge
may occur when the leading follicle is much smaller, as shown by the
example in Figure 2, where the largest follicle diameter was only 14 mm,
but with plasma oestradiol at supranormal concentrations.

The surge of LH may demonstrate normal or attenuated concentration
profiles[7] to which the follicles may respond in a number of ways, some of
which parallel those observed in the rabbit experiments above. This
combination of abnormal circumstances results in a number of problems
(at the levels of the oocyte, ovary and endometrium) of which only a few
have been clearly identified.

Ovulation induction in polycystic ovary syndrome

Ovulation induction in infertile patients with polycystic ovary syndrome
(PCOS) and oligomenorrhoea receiving human menopausal gonadotrophin (hMG) yields a pregnancy rate below that of patients with
hypothalamic amenorrhoea treated identically[8]. A comparison of the
hormone profiles of cycles resulting in pregnancy ($n = 13$) with 13 cycles
treated simultaneously but failing to yield pregnancy is shown in Figure 3.
The profile of progesterone in the conception group describes a curve
similar to that of the normal cycle, while that of the non-conception
group was significantly elevated over the 4 days preceding hCG administration. The mean LH profiles show that levels in the non-conception
group were above the normal range throughout the period of observation
while the conception cycle data showed supranormal levels only on the
day of hCG treatment. The mean LH data shown here obscure a series of
individual LH surges observed throughout the late follicular phase in the

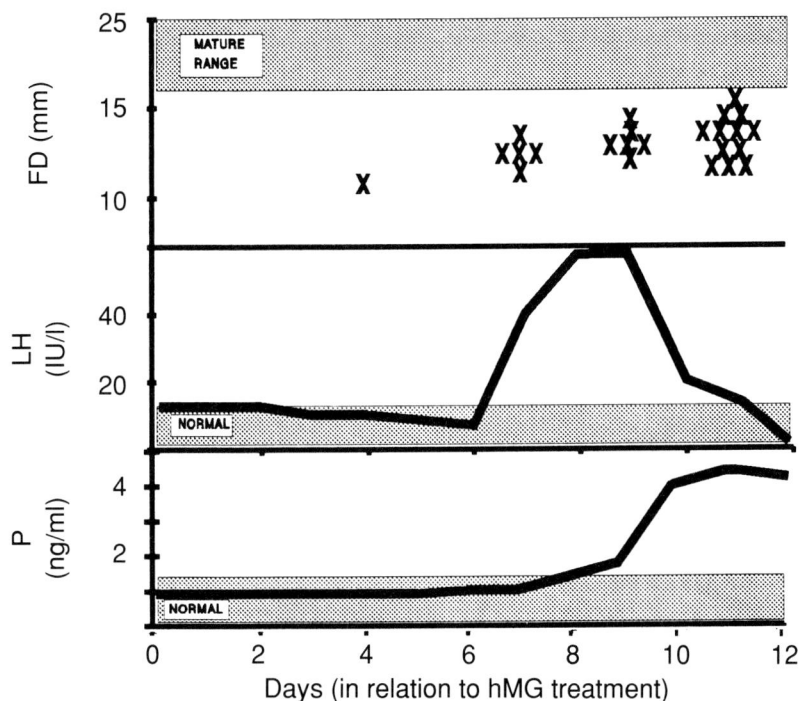

Figure 2 Profiles of follicular diameter (FD), luteinizing hormone (LH) and progesterone (P) in a patient showing premature luteinization. The stippled background represents the normal cycle follicular phase hormone data, and normal, day 0 follicle diameters

non-conception group. Luteinizing hormone surges seen in the conception group were restricted to the final 24 h before treatment with hCG.

These data imply that premature luteinization results in a reduced fecundity in stimulated cycles.

EVIDENCE FROM IVF

The resource commitment for IVF programmes has demanded greater understanding of the circumstances and consequences of premature

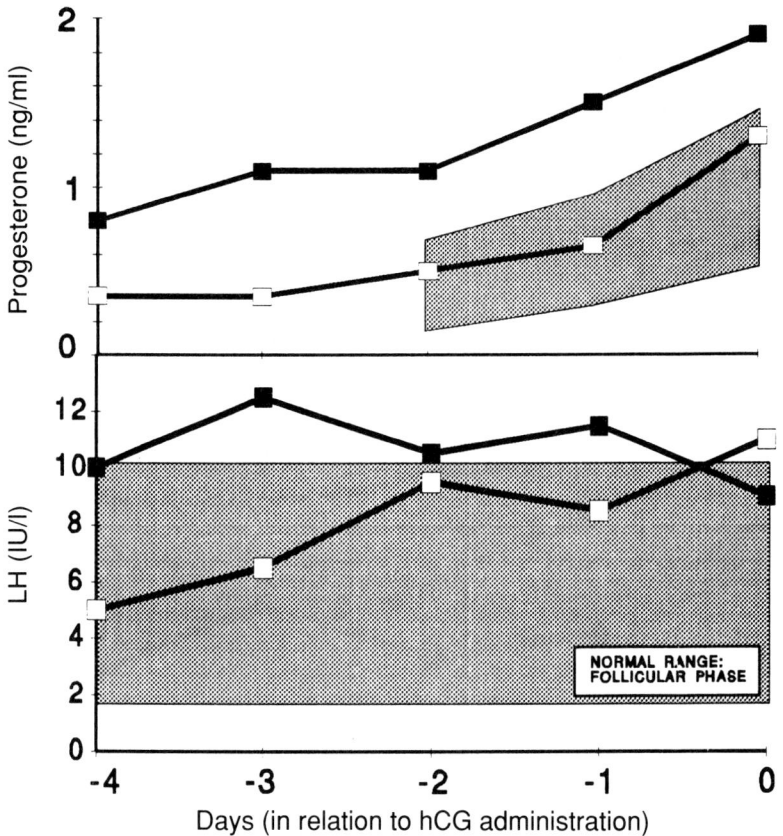

Figure 3 Mean hormone profiles in patients with polycystic ovaries treated with exogenous gonadotrophins (human menopausal gonadotrophin, hMG). The conception cycles (□, $n = 13$) are shown compared with simultaneously treated non-conception cycles (■, $n = 13$). The stippled areas represent normal follicular phase luteinizing hormone (LH) concentrations and normal conception cycle progesterone data relative to the LH peak

luteinization, and it has also provided a substantial amount of data unavailable through ovulation induction.

Incidence and complications

All cycles of multiple follicular development do not result in premature luteinization, but in the absence of GnRH-a, rearrangement of timing for oocyte retrieval or cycle cancellation occurred in up to 40% of cycle starts irrespective of the method of stimulation[9]. The LH surge with luteinization (progesterone > 1.5 ng/ml) has been recorded to occur before the identification of a follicle with a diameter ≥ 17 mm in 40% of cycles treated with hMG alone and in 51% before a 20 mm follicle was seen[10]. However, the value judgement of when to adapt clinical procedures depends on other factors, such as the number of medium-sized follicles present and also the access to clinical facilities.

When the luteinization was observed at a time when follicles were relatively small (< 17 mm) an increased incidence of oocyte retention within the follicle was recorded, indicating that normal ovulatory processes were not fully developed until the follicle attained normal preovulatory dimensions[11]. The implications of this extend beyond mere release of the oocyte, since the abnormal environment may have deleterious effects upon oocyte/embryo viability, and use of these oocytes in an IVF programme may result in a lower implantation rate. This parallels the rabbit experiment described previously. The ability to delay the luteinization signal until a suitable number of follicles are of mature dimensions is crucial to improved IVF results. This is the critical reason for the enthusiastic adoption of the use of GnRH analogues in IVF programmes.

Oocyte postmaturity and GnRH analogues

A comparison of IVF cycles treated with hMG alone, or combined with long-course GnRH-a, revealed that postmature oocytes (PMOs) were exclusively found in hMG-alone cycles. They were observed in 10 cycles out of 51 oocyte retrievals effected after 75 cycle starts. They were identified by their dark contracted corona radiata, which obscures the oocyte from view and were associated with a disintegrated, friable cumulus mass. They were usually found with apparently normal oocytes in the same cohort, but the cycles demonstrated significantly reduced fertilization and pregnancy rates[11].

Retrospective hormone analyses of cycles with PMOs, monitored in plasma samples taken twice daily, revealed that every cycle demonstrated a surge of LH during the 5 days prior to, and more than 24 h before the hCG was administered (Figure 4). They also showed that in all cases the LH surge was attenuated and short-lived: such that their validity was

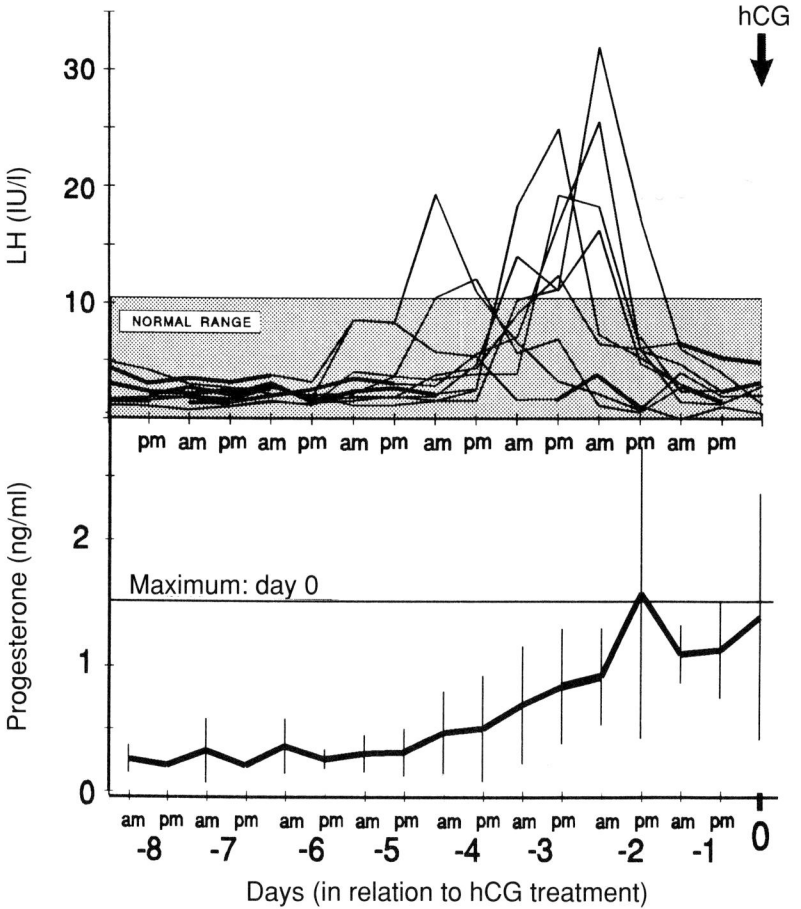

Figure 4 Luteinizing hormone (LH) and progesterone profiles in women treated with menopausal gonadotrophin and who produced postmature oocytes. The patients were sampled twice daily and the stippled background shows the normal follicular phase LH concentration

only confirmed by the twice-daily sampling employed. The moderate elevations in progesterone, which were close to the upper limit, were insufficient to demand cycle cancellation for which the criterion was a progesterone level > 1.5 ng/ml in two successive samples. The low pregnancy potential of these cycles may be due to effects at both the follicular/oocyte level and/or at the endometrial level, since premature elevations in progesterone would influence endometrial receptivity for non-synchronized embryos.

These data indicate that there are limitations on the ability of plasma concentrations of progesterone to detect subnormal luteinization as seen in the PMO cycles, and that the attenuated LH surges seen in stimulated cycles do interfere with clinical processes in an unpredictable manner, best avoided using GnRH-a-induced suppression of LH. The data do not support the hypothesis that high basal LH values are the cause of oocyte/embryo developmental errors[12] – only that the inappropriate LH surge is responsible for these elements.

EXPERIMENTAL PREMATURE LUTEINIZATION

Short-course GnRH-a protocols usually employ the agonistic flare effect of initial GnRH-a stimulation in the early follicular phase to initiate recruitment. The flare effect is characterized by increases in FSH concentrations which can be used to stimulate initial follicular development, but it also produces an increase of LH which may be used to simulate an LH surge. In this study, a fixed protocol was used to investigate the effects of this LH surge in 20 IVF cycles in which follicular recruitment and development had been initiated by administration of FSH (Metrodin®, Serono UK Ltd; 150 IU/day) on cycle days 1–5. The GnRH-a was administered from day 5 and follicular development was continued with daily hMG (225 IU/day).

Figure 5 shows the mean hormone profiles in these cycles which demonstrate that the GnRH-a induced premature luteinization with increasing progesterone concentrations through the late follicular phase as hCG administration (day 0) approached. This was seen despite declining LH concentrations. Table 1 compares the outcome of these cycles with 20 long-course control cycles (GnRH-a initiated in the previous luteal phase) and it shows the catalogue of follicle, oocyte and embryo problems found in the experimental cycles. There was an increased incidence of ovarian cysts and follicles whose fluids showed no granulosa cells, along with a reduced yield of oocytes, and a significantly reduced fertilization rate in these cycles. These abnormalities were not all associated with oocyte postmaturity, but the reduced rates of normal fertilization indicated

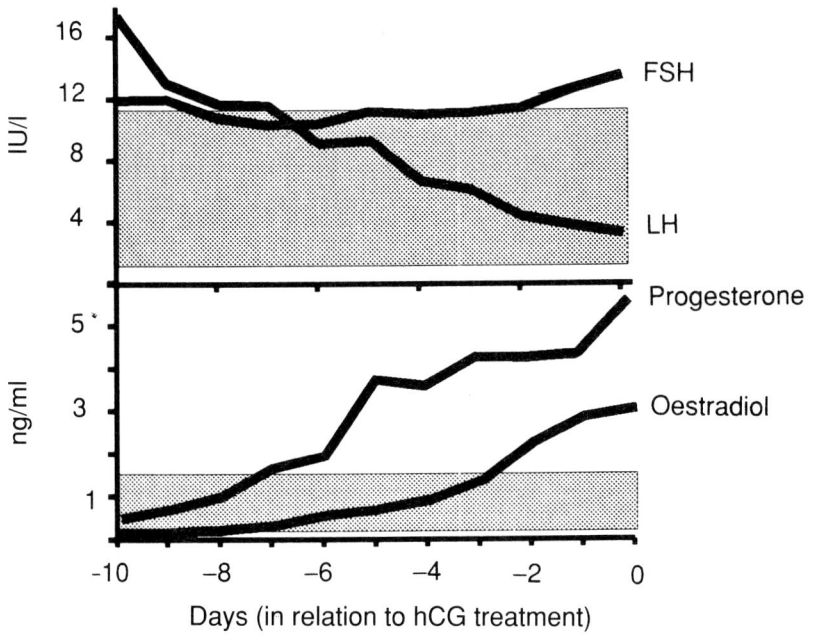

Figure 5 Mean hormone profiles in patients ($n = 20$) with experimentally induced premature luteinization with data relative to the administration of human chorionic gonadotrophin (hCG). The stippled backgrounds show the normal follicular phase luteinizing hormone (LH) concentration and the normal day 0 progesterone concentration. FSH, follicle stimulating hormone

that abnormal oocyte development was common. These profiles confirm that the premature LH surge compromises oocyte development.

SUMMARY

Premature luteinization has been shown to occur in a high percentage of cycles of induced multiple follicular development. The LH surge may occur when the dimensions of the leading follicles are immature and, although the surge itself may be attenuated, reduced cycle fecundity results. Animal experiments have shown that premature luteinization with an attenuated luteinizing signal can induce abnormalities of follicle

Table 1 Ovarian and embryological effects of experimentally induced premature luteinization ($n = 20$ cases) compared with long-course GnRH-a-treated controls ($n = 20$ cases). FD, follicle diameter

	Control	*Experimental*
Mature follicles	6.4	5.9
Cysts (cases with FD > 25 mm)	2	11★
Oocytes	8.1	5.0
Fertilization/division (%)	70	49★
Postmaturity (cases)	0	2 (10%)
Acellular follicles (cases)	0	2 (10%)

★ $p < 0.05$

differentiation and/or corpus luteum formation, and oocyte nuclear maturation. Similar observations have now been recorded in humans leading to reduced fertilization/division rates and fecundity.

In cycles treated with hMG alone, oocyte postmaturity was universally associated with a premature LH surge, reduced fertilization rates and fecundity, but this is by no means the only follicle/oocyte abnormality associated with premature luteinization.

The issue of asynchronous endometrial development, due to early rises of progesterone, has not been discussed, but it is a further theoretical source of reduced fecundity in cycles showing premature luteinization.

Combined GnRH-analogue and hMG therapy eliminates the problems of premature luteinization and its associated abnormalities, and thereby eliminates the need for comprehensive cycle monitoring. The benefits of this form of treatment are well-established in IVF programmes, although it is not so universally employed in programmes of follicular stimulation for ovulation induction with or without intrauterine insemination, even though the same criteria apply.

The data presented above indicate that combined GnRH-analogue therapy should be employed in all cases of ovulation induction in patients with intact pituitary function.

REFERENCES

1. Gemzell, C.A., Kemman, E. and Jones, J.R. (1978). Premature ovulation during administration of human menopausal gonadotropins in non-ovulatory women. *Infertility*, **1**, 1–10

2. Hamilton, M.P.R., Fleming, R., Coutts, J.R.T., Macnaughton, M.C. and Whitfield, C.R. (1990). Luteal phase deficiency: ultrasonic and biochemical insights into pathogenesis. *Br. J. Obstet. Gynaecol.*, **97**, 569–75

3. Marik, J. and Hulka, J. (1978). Luteinized unruptured follicle syndrome: a subtle cause of infertility. *Fertil. Steril.*, **29**, 270–4

4. Hamilton, M.P.R., Fleming, R., Coutts, J.R.T., Macnaughton, M.C. and Whitfield, C.R. (1990). Luteal cysts and unexplained infertility: biochemical and ultrasonic evaluation. *Fertil. Steril.*, **54**, 32–7

5. Hackeloer, B.J., Fleming, R., Robinson, H.P., Adams, A.H. and Coutts, J.R.T. (1979). Correlation of ultrasonic and endocrinological assessment of follicular development. *Am. J. Obstet. Gynecol.*, **135**, 122–8

6. Messinis, I.E., Templeton, A. and Baird, D.T. (1986). Relationships between the characteristics of endogenous luteinizing hormone surge and the degree of ovarian hyperstimulation during superovulation induction in women. *Clin. Endocrinol.*, **25**, 393–400

7. Lunenfeld, B. and Insler, V. (1978). *Diagnosis and Treatment of Functional Infertility*, pp. 61–89. (Berlin: Gross Verlag)

8. Fleming, R. and Coutts, J.R.T. (1990). Induction of multiple follicular development for IVF. *Br. Med. Bull.*, **46**, 596–615

9. Fleming, R. and Coutts, J.R.T. (1986). Induction of multiple follicular growth in normally menstruating women with endogenous gonadotropin suppression. *Fertil. Steril.*, **45**, 226–30

10. Stanger, J.D. and Yovich, J.L. (1984). Failure of human oocyte release at ovulation. *Fertil. Steril.*, **41**, 827–32

11. Jamieson, M.E. (1992). *Biological Influences On The Cytogenetics of Early Human Development: An IVF Study*. (PhD. thesis, University Of Glasgow).

12. Regan, L., Owen, E.J. and Jacobs, H.S. (1990). Hypersecretion of luteinizing hormone, infertility, and miscarriage. *Lancet*, **336**, 1141–4

12

Corpus luteal insufficiency

M.G. Hammond

ASSESSMENT OF CORPUS LUTEAL FUNCTION

Normal corpus luteal function is dependent on normal follicle growth, adequate numbers of granulosa-lutein cells and the occurrence of the luteinizing hormone (LH) surge. Progesterone production confirms luteinization of the follicle but not ovum release. The three primary features for documenting progesterone production and corpus luteum formation are basal body temperature, endometrial biopsy and serum progesterone levels[1].

Basal body temperature

BBT is a daily record of oral temperature recorded by the patient on waking. Progesterone secretion produces a 0.5–1°F temperature shift. Biphasic temperature charts are said to be diagnostic of ovulation. Basal body temperature is most useful to pinpoint the approximate time of ovulation and to assess the length of the luteal phase. It does not confirm adequate luteal function. The precise length of the normal luteal phase is reported to be a mean of 14.13 days from the LH surge. Temperature elevation over less than 10 days is abnormal[2].

Endometrial biopsy

Endometrial biopsy has been used to confirm ovulation. Noyes and colleagues established the criteria for dating the normal late luteal endometrial biopsy[3]. The technique is a time-consuming and expensive means to confirm ovulation. Jacobson and Marshall[4] noted significantly reduced conception rates in biopsied cycles of patients treated with human menopausal gonadotrophin (hMG) and human chorionic gonadotrophin (hCG). The recent introduction of the flexible plastic cannula has reduced the morbidity of the procedure. If endometrial biopsies are to be utilized both to confirm ovulation and assess luteal function, Jones[5] recommends that they be obtained immediately prior to the expected menses (day 26 of a 28-day cycle) and assessed by the criteria of Noyes and colleagues[3]. If abnormal, the biopsy should be repeated in a subsequent cycle. Biopsies are abnormal if they are asynchronous by more than 2 days from the cycle day calculated retrospectively from the subsequent menstrual period. Recent studies suggest timing from the LH surge may be more accurate[6]. Abstinence or barrier contraception can be used if damage to a conceptus is feared.

Serum progesterone

The use of luteal phase serum progesterone to confirm ovulation is becoming increasingly common. Physician visits are not required, and numerical values which are reproducible are obtained from different laboratories. Serum progesterone levels of ≥ 3 ng/ml (10.2 nmol/l) were noted in the mid-luteal phase of regularly cycling women by Israel and co-workers[7]. Others[8-10] have noted serum progesterone levels of ≥ 9–10 ng/ml (30.6–34 nmol/l) in spontaneous conception cycles monitored in the mid-luteal phase[8]. Hull and colleagues[10] and Hammond and co-workers[8] have observed that higher progesterone levels (we use 15 ng/ml (51 pmol/l)) are evident in ovulatory clomiphene treatment cycles. Samples should be obtained 5–10 days prior to menses. Progesterone levels in spontaneous cycles usually reach a plateau from days 19 to 23, giving peak values during this time. We obtain progesterone on day 21 of clomiphene treatment cycles or 1 week after hCG injection.

PROBLEMS IN EVALUATING LUTEAL PHASE INSUFFICIENCY

Many reports have questioned the reliability of single or even multiple progesterone assays and endometrial biopsies in characterizing the average cycle of the individual patient. These critiques take two forms. One is that normal women have occasional abnormal cycles (frequency unknown). Thorough and accurate evaluation of one of these cycles would not be predictive of the patients' characteristic ovulatory function. Assessment of normal populations of women with endometrial biopsies has revealed abnormal endometrial development in 20% of cycles. On the basis of probability one could then expect two serial abnormal biopsies in 4% of normal women[11]. Similarly, low progesterone levels might be noted in sporadic cycles of normally ovulating women. The second criticism is that the laboratory assessments have inherent pitfalls which may make isolated values meaningless.

It has been observed that progesterone is released in a pulsatile fashion[12]. Recent observations in our centre have revealed that by the mid-luteal phase the frequency of pulsatile progesterone production is reduced, and a pattern similar to the diurnal pattern of cortisol production is noted[13].

Variation in progesterone levels has been approached by frequent measurement of salivary progesterone[14] which is less invasive. Multiple samples can be stored and run simultaneously. Serial sampling gives a more representative picture of corpus luteum function throughout the cycle.

Difficulties in interpretation of the endometrial biopsy and individual variation in dating of samples has been well documented. Noyes and Haman[11] found a 63% discrepancy between the results of two pathologists. Additional refinement of histologic criteria has also been suggested[15,16]. Perhaps it would be more useful to develop quantitative tests of endometrial maturation, such as the measurement of the production of prolactin[17], or assays for serum levels of specific endometrial secretory proteins to evaluate endometrium function.

Optimal luteal phase assessment may require the measurement of basal body temperature, serum progesterone and endometrial biopsy. Shangold and colleagues[18] have demonstrated a poor correlation between these parameters in a series of 34 infertile women, suggesting that all three should be measured to evaluate the luteal phase adequately.

TREATMENT OF LUTEAL PHASE DEFECTS

Progesterone

The use of progesterone for the treatment of luteal-phase defects was first suggested by Jones[5]. Patients diagnosed by two out-of-phase endometrial biopsies are treated with progesterone supplementation by suppository or intramuscular injection. The drug acts by correcting for an inadequate production of progesterone by the corpus luteum or by correcting a local endometrial progesterone deficiency, thus normalizing late luteal biopsies.

Progesterone is usually administered by mucosal absorption per vagina or rectum or by intramuscular injection. Synthetic progestins are avoided. Several new oral preparations using micronized progesterone are under investigation[19], but variable absorption has been reported. Therapy is begun 3 days after the temperature shift of basal body temperature and continued until menses. If pregnancy occurs, supplementation is continued to 10 weeks. 17-hydroxyprogesterone caproate (250 mg) can be substituted weekly at 6 weeks' gestation.

Progesterone suppositories, 25 mg twice a day, or progesterone in oil, 12.5 mg/day, intramuscularly, may be utilized. Rarely, the onset of menstruation may be delayed by the medication. If pregnancy testing is negative, the drug should be discontinued and menses should ensue. The drug is then reinstituted after the next ovulation. Therapy is monitored by endometrial biopsy during the treatment cycle and doses are increased until the defect is corrected. Therapy is then continued for six cycles.

Many small series have been reported with uncorrected pregnancy rates of about 50%. A recent study of 54 women by Maxson and co-workers reported a 43% pregnancy rate[20]. No life-table analysis has been reported; however, the mean cycle of conception is 2.4 months[21]. Wentz and colleagues have recently reviewed their experience with the evaluation and therapy of luteal phase inadequacy and provide an excellent discussion of the impact of this disorder on fertility[22]. Balasch and colleagues[23] have recently surveyed the usefulness of endometrial biopsy for luteal phase evaluation in infertility. They concluded that the diagnosis represents a chance event, abnormal biopsies do not affect outcome of pregnancy and treatment does not improve pregnancy rates.

Table 1 Treatment plan for improved luteal function with clomiphene

Monitoring of therapy
- (1) Basal body temperature charting
- (2) Treat on days 3–7 of cycle
- (3) Measure serum progesterone on day 21
- (4) If progesterone is between 2 and 15 ng/ml, increase dose
- (5) If progesterone is at least 15 ng/ml and menses occur, continue this dose. If no menses ensue, obtain serum pregnancy test
- (6) Always obtain a serum pregnancy test prior to medroxyprogesterone or clomiphene treatment if there is any question of pregnancy

Dosage plan
- (1) Clomiphene, 50 mg, for one cycle if progesterone is low
- (2) Increase to 100 mg for one cycle if progesterone is low
- (3) Add hCG, 10 000 units, on day 15 or 16 if progesterone is low
- (4) Increase clomiphene to 150 mg and hCG on day 15 to 16
- (5) If progesterone is ≥ 12 ng/ml, continue three cycles. If progesterone is < 10 ng/ml, proceed to menotropins
- (6) Use high-dose clomiphene (> 150 mg/day) only for patients weighing 82 kg (180 lb) or more

Clomiphene citrate

The mode of action of clomiphene citrate remains speculative, but increasing evidence supports the proposal that it interferes with hypothalamic or pituitary cytosol oestrogen receptor binding or replenishment, blocking oestrogen feedback, which leads to increasing levels of serum gonadotrophins. This results in improved folliculogenesis and, frequently, to increased follicle and corpus luteal numbers and increased serum progesterone levels. Guzick and Zeleznik[24] propose, from ultrasound analysis of clomiphene-treated cycles, that the correction of luteal phase abnormalities is not due to improved folliculogenesis but to increased follicle number, because only two of eight patients with one follicle had normalization of their biopsies whilst eight of ten patients with more than one follicle had normal biopsies.

Clomiphene citrate administration begins on the 3rd–5th day after the onset of spontaneous menses (Table 1). Therapy is initiated at 50 mg per day for 5 days; the response is assessed by serum progesterone or endometrial biopsy during the luteal phase. If the luteal phase is inadequate, the dosage is increased to 100 mg daily for 5 days in the subsequent cycle. Dosage may be increased in this fashion up to a total of 150 mg/day.

The addition of hCG (10 000 IU administered intramuscularly on day 14 or 15 of the cycle) improves results. This drug is useful in patients who develop adequate folliculogenesis but mount an inadequate endogenous midcycle LH peak or show persistent low progesterone values. Repeat injections (5000 IU) 1 week later may stimulate the corpus luteum and sustain the luteal phase. Ultrasound monitoring of follicle diameter may be used to time hCG injection.

LUTEAL PHASE INSUFFICIENCY DURING OVULATION INDUCTION

Clomiphene citrate

Multiple authors have addressed the role of clomiphene citrate in the development of the endometrium. Clomiphene is an anti-oestrogen and early authors suggested that ovulatory cycles on clomiphene citrate had an increased incidence of luteal inadequacy[25,26]. Lamb and co-workers[27] addressed this question by performing serial endometrial biopsies on women ovulating with clomiphene citrate treatment and showed no evidence of luteal insufficiency. Birkenfeld and colleagues[28] have recently reviewed studies up to 1986. Many investigators have used endometrial biopsies to assess luteal phase function, but very few have confirmed normal ovulation using progesterone levels in the same studies.

Recently, new technologies have been brought to bear on the study of the effect of clomiphene citrate on the endometrium. Oestrogen and progesterone receptors have been investigated in normal women prior to and during clomiphene citrate therapy[29]. The authors have demonstrated no change in the concentrations or binding constants for oestrogen or progesterone receptors in the mid-luteal phase of normal women treated with up to 150 mg clomiphene citrate.

Ultrasound techniques have also been utilized. Eden and colleagues[30] used serial ultrasound to study women in both a natural cycle and a clomiphene citrate-induced cycle. Despite higher oestrogen levels in the preovulatory phase, they noted a decrease in the normal uterine volume and endometrial thickening seen in the preovulatory period. However, Grunfeld and co-workers[31] have demonstrated, with endovaginal ultrasound, that the thickness of the endometrium is not as helpful as a homogeneous hypoechoic pattern for predicting normal histology.

Treatment for luteal deficiency detected in the endometrium of patients on clomiphene citrate ovulation induction should involve careful analysis to determine that adequate folliculogenesis is occurring and that post-ovulatory serum progesterones are normal. If not, the dose of clomiphene citrate could be increased or hCG could be administered. An ultrasound evaluation of follicular development and rupture is important. In the presence of normal folliculogenesis and follicle rupture, additional progesterone supplementation, either by suppository or injection, might be indicated or the patient might be transferred to gonadotrophin ovulation induction therapy.

Human menopausal gonadotrophins

Studies of the effect of human menopausal gonadotrophin (hMG) on the luteal phase are less common than those with clomiphene. From close analysis of folliculogenesis during hMG therapy, the assumption has been made that the luteal phase will be normal. Olson and colleagues[32] initially reported on shortened luteal phases following ovulation induction. They noticed a correlation with oestradiol levels < 200 or > 2000 pg/ml. These findings have been confirmed by other investigators.

Spontaneous abortion rates have also been reported to be higher than normal in patients receiving hMG therapy. This higher abortion rate, 41.7% versus 22.8%, appears to be associated with the World Health Organization (WHO) group II, those with chronic anovulation and/or polycystic ovarian disease. Abortion rates in patients with hypo-oestrogenism appear to be the same as normal[33].

There have been few controlled studies evaluating the appropriate therapy of the luteal phase in patients receiving gonadotrophin therapy. Some authors have recommended serial hCG injections, varying from two injections (on the day of ovulation and 1 week later) to three or even four injections every 2–3 days throughout the early part of the luteal phase. In addition, some centres supplement with progesterone injections,

Table 2 Luteal support during ovulation induction with human chorionic gonadotrophin (hCG) (data from ref. 34)

Group	No hCG (%)	4500 IU hCG (%)
Luteal phase defect		
WHO I	21	1.8
WHO II	9	5.2
Pregnancy		
WHO I	9	31.9
WHO II	18	21

25–50 mg daily, similarly to their *in vitro* fertilization (IVF) programmes. Messinis and colleagues[34] recently evaluated the role of multiple hCG injections in patients receiving hMG, dividing these patients by WHO classifications. In 205 out of 341 treatment courses, additional hCG was administered. In WHO group I the incidence of luteal phase defects was lower and the pregnancy rate higher in cycles with extra hCG. In WHO group II there was no such difference after supplemental hCG. The abortion rate was the same after cycles with or without additional hCG (Table 2).

GONADOTROPHINS AND GnRH AGONISTS

The introduction of GnRH agonists (GnRH-a) in ovulation induction protocols has led to new concerns in regard to the luteal phase of the cycle. There has been considerable debate over the role of LH during the luteal phase. Various studies evaluated the LH dependency of the corpus luteum, utilizing pituitary surgery, antibodies to LH and GnRH-a [35,36]. The consensus seems to be that an LH-like activity is required for normal corpus luteal function. Studies within an IVF programme[37] demonstrated that, following the discontinuation of agonists on the day of hCG administration, the LH level remained undetectable for the next 10 days, and lower than that in a control group for 6 further days, suggesting that the pituitary gland does not recover from suppression during the luteal phase of the treatment cycle. Multiple treatment options are available for correction of this defect, including progesterone supplementation by injection or suppository, or serial hCG injections.

LUTEAL INSUFFICIENCY ASSOCIATED WITH *IN VITRO* FERTILIZATION CYCLES

Early in the study of *in vitro* fertilization, investigators became concerned about corpus luteal function because of the aspiration of granulosa cells at the time of egg retrieval. Because of low implantation rates using IVF, attention was focused on endometrial development. Endometrial biopsies were performed during non-transfer cycles.

In the early years of IVF, when ovulation induction agents were used alone, attention was focused on the effect of clomiphene or gonado-trophins on luteal function. Following the introduction of gonadotrophin releasing hormone agonists, attention then turned to the effect of the agonist as well.

In the absence of agonists, clomiphene citrate and human menopausal gonadotrophins have similar effects in IVF and anovulatory patients, leading to endometrial development and the short luteal phase. Studies have demonstrated that there is both delayed and advanced endometrial development. Many centres have chosen supplementation of the luteal phase with progesterone. Meta-analysis of these findings failed to show a statistically significant improvement[38]. Trounson and co-workers[39] reported no difference in pregnancy rates between patients treated with progesterone, hCG or placebo. Thus, in IVF cycles in which GnRH agonists are not utilized, there is no support for improved pregnancy rates with progesterone support.

In the presence of GnRH agonists, concern is warranted because of the slow recovery of the pituitary gland and the rise in LH levels for corpus luteal support. Several studies have demonstrated an increased pregnancy rate when hCG supplementation is added to GnRH-a human menopausal gonadotrophin IVF cycles[40,41]. The various recommended supplementation regimens are shown in Table 3.

Table 3 Luteal support during *in vitro* fertilization with GnRH agonists. hCG, human chorionic gonadotrophin

Progesterone suppositories, 100–400 mg/day
Oral progesterone, 200 mg twice daily
Progesterone, 50–100 mg/day
hCG, 2000 IU, day 4, 8, 12
hCG, 1500 IU, day 2, 4, 6

CONCLUSION

Multiple means are available for the assessment of luteal function in the infertile woman. The role of luteal phase deficiency as a cause of significant infertility in spontaneous ovulating women has been questioned.

Luteal phase dysfunction during ovulation induction is more common. High preovulatory oestradiol levels and the use of GnRH-a increase the risk of luteal abnormalities. There is some evidence to support improved pregnancy rates with luteal phase supplementation in gonadotrophin ovulation and GnRH agonist-treated patients.

REFERENCES

1. McNeely, M.J. and Soules, M.R. (1988). Diagnosis of luteal phase deficiency: a critical review. *Fertil. Steril.*, **50**, 1–15
2. Lenton, E.A., Landgren, B.-M. and Sexton, L. (1984). Normal variations in the length of the luteal phase: identification of the short luteal phase. *Br. J. Obstet. Gynaecol.*, **91**, 685–9
3. Noyes, R.W., Hertig, A.T. and Rock, J. (1950). Dating the endometrial biopsy. *Fertil. Steril.*, **1**, 3–25
4. Jacobson, A.J. and Marshall, J.R. (1980). Detrimental effect of endometrial biopsies on pregnancy rate following human menopausal gonadotropin/human chorionic gonadotropin-induced ovulation. *Fertil. Steril.*, **33**, 602–4
5. Jones, G.S. (1976). The luteal phase defect. *Fertil. Steril.*, **27**, 351–6
6. Li, T.C., Rogers, A.W., Lenton, E.A., Dochery, P. and Cooke, I. (1987). A comparison between two methods of chronological dating of human endometrial biopsies during the luteal phases, and their correlation with histologic dating. *Fertil. Steril.*, **48**, 928–32

7. Israel, R., Mishell, D.R., Stone, S.C., Thorneycroft, I.H. and Moyer, D.L. (1972). Single luteal phase serum progesterone assay as an indication of ovulation. *Am. J. Obstet. Gynecol.*, **112**, 1043–6

8. Hammond, M.G. and Talbert, L.M. (1982). Clomiphene citrate therapy of infertile women with low luteal phase progesterone levels. *Obstet. Gynecol.*, **59**, 275–9

9. Swyer, G.I.M., Radwanska, E. and McGarrigle, H.H.G. (1975). Plasma estradiol and progesterone estimation for the monitoring of induction of ovulation with clomiphene and chorionic gonadotropin. *Br. J. Obstet. Gynaecol.*, **82**, 794–804

10. Hull, M.G.R., Savage, P.E., Bromham, D.R., Ismail, A.A.A. and Morris, A.F. (1982). The value of a single serum progesterone measurement in the midluteal phase as a criterion of a potentially fertile cycle ("ovulation") derived from treated and untreated conception cycles. *Fertil. Steril.*, **37**, 355–60

11. Noyes, R.W. and Haman, J.O. (1953). Accuracy of endometrial dating: correlation of endometrial dating with basal body temperature and menses. *Fertil. Steril.*, **4**, 504–17

12. Filicori, M., Butler, J.P. and Crowly, W. Jr (1984). Neuroendocrine regulation of the corpus luteum in the human: Evidence of pulsatile progesterone secretion. *J. Clin. Invest.*, **73**, 1638–47

13. Syrop, C.H. and Hammond, M.G. (1987). Diurnal variations in midluteal serum progesterone measurements. *Fertil. Steril.*, **47**, 67–70

14. Choe, J.K, Khan-Dawood, F.S. and Daywood, M.Y. (1983). Progesterone and estradiol in the saliva and plasma during the menstrual cycle. *Am. J. Obstet. Gynecol.*, **147**, 557–62

15. Johannisson, E., Parker, R.A., Landgren, B.M. and Diczfalusy, E. (1982). Morphometric analysis of human endometrium in relation to peripheral hormone levels. *Fertil. Steril.*, **38**, 564–71

16. Li, T.C., Dochery, P., Rogers, A.W. and Cooke, I.D. (1989). How precise is histologic dating of endometrium using the standard dating criteria? *Fertil. Steril.*, **51**, 759–63

17. Daly, D.C., Maslar, I.A., Rosenberg, S.M., Tohan, N. and Riddick, D.H. (1981). Prolactin production by luteal phase defect endometrium. *Am. J. Obstet. Gynecol.*, **140**, 587–91

18. Shangold, M., Berkeley, A. and Gray, J. (1983). Both midluteal serum progesterone levels and late luteal endometrial histology should be assessed in all infertile women. *Fertil. Steril.*, **40**, 627–30

19. Padwich, M.L., Endicott, J., Matson, C. and Whitehead, M.I. (1986). Absorption and metabolism of oral progesterone when administered twice daily. *Fertil. Steril.*, **46**, 402–7

20. Maxson, W.S., Wentz, A.C. and Herbert, C.M., (1984). Outcome of progesterone therapy on luteal phase inadequacy. *Fertil. Steril.*, **41**, 856–62
21. Daly, D.C., Walters, C.A., Soto Albors, C. *et al.* (1982). Multiple therapeutic modalities improve pregnancy rates in luteal phase defects. *Fertil. Steril.*, **39**, 393a
22. Wentz, A.C., Kossoy, L.R. and Parker, R.A. (1990). The impact of luteal phase inadequacy in an infertile population. *Am. J. Obstet. Gynecol.*, **162**, 937–45
23. Balasch, J., Fabregues, F., Sreus, M. and Vanrell, J.A. (1992). The usefulness of endometrial biopsy for luteal phase evaluation in infertility. *Hum. Reprod.*, **7**, 973–7
24. Guzick, D.S. and Zeleznik, A. (1990). Efficacy of clomiphene citrate in the treatment of luteal phase deficiency: quantity versus quality of preovulatory follicles. *Fertil. Steril.*, **54**, 206–10
25. Van Hall, E.V. and Mastboom, J.L. (1969). Luteal phase insufficiency in patients treated with clomiphene. *Am. J. Obstet. Gynecol.*, **103**, 165–71
26. Jones, G.S., Maffezzoli, R.D. and Strottca, *et al.* (1970). Pathophysiology of reproductive failure after clomiphene induced ovulation. *Am. J. Obstet. Gynecol.*, **108**, 847–67
27. Lamb, E.J., Colliflower, W.W. and Williams, J.W. (1972). Endometrial histology and conception rates after clomiphene citrate. *Obstet. Gynecol.*, **39**, 389–96
28. Birkenfeld, A., Beier, H.M. and Schenker, J.G. (1986). The effect of clomiphene citrate on early embryonic development, endometrium and implantation. *Reproduction*, **1**, 387–95
29. Hecht, B.R., Khan-Dawood, F.S. and Dawood, M.Y. (1989). Peri-implantation phase endometrium estrogen and progesterone receptors: effect of ovulation induction with clomiphene citrate. *Am. J. Obstet. Gynecol.*, **161**, 1688–93
30. Eden, J.A., Place, J., Carter, G.D., Jones, J., Alaghband-Zadeh, J. and Pawson, M.E. (1989). *Obstet. Gynecol.*, **73**, 187–90
31. Grunfeld, L., Walker, B., Bergh, P.A., Sandler, B., Hofmann, G. and Navot, D. (1991). High-resolution endovaginal ultrasonography of the endometrium: a noninvasive test for endometrial adequacy. *Obstet. Gynecol.*, **78**, 200–204
32. Olson, J.L. and Rebar, R.W., Schreiber, J.R., Vaitukaitis, J.L. (1983). Shortened luteal phase after ovulation induction with human menopausal gonadotropin and human chorionic gonadotropin. *Fertil. Steril.*, **39**, 284–91
33. Oelsner, G., Serr, D.M., Mashiach, S., Blankstein, J., Snyder, M. and Lunenfeld, B. (1978). The study of induction of ovulation with menotropins: analysis of results of 1897 treatment cycles. *Fertil. Steril.*, **30**, 538–44

34. Messinis, I.E., Bergh, T. and Wide, L., (1988). The importance of human chorionic gonadotropin support of the corpus luteum during human gonadotropin therapy in women with anovulatory infertility. *Fertil. Steril.*, **50**, 31–5

35. Asch, R.H., Abou-Samra, M., Braunstein, G.D. and Pauerstein, C.J. (1982). Luteal function in hypophysectomized Rhesus monkeys. *J. Clin. Endocrinol. Metab.*, **55**, 154–61

36. Hutchison, J.S. and Zeleznik, A.J. (1984). The Rhesus monkey corpus luteum is dependent on pituitary gonadotropin secretion throughout the luteal phase of the menstrual cycle. *Endocrinology*, **115**, 1780–6

37. Balasch, J. and Van Rell, J.A. (1987). Corpus luteum insufficiency and infertility, a matter of controversy. *Hum. Reprod.*, **2**, 557–67

38. Daya, S. (1988). Efficacy of progesterone support in the luteal phase following *in vitro* fertilization and embryo transfer: meta-analysis of clinical trials. *Hum. Reprod.*, **3**, 731–4

39. Trounson, A., Howlett, D., Rogers, R. and Hoppen, H.O. (1986). The effect of progesterone supplementation around the time of oocyte recovery in patients supraovulated for *in vitro* fertilization. *Fertil. Steril.*, **45**, 532–5

40. Belaisch-Allart, J., Dmouzon, J., Lapousterle, C. and Mayer, M. (1990). The effect of hCG supplementation after combined GnRH agonist hMG treatment in an IVF program. *Hum. Reprod.*, **5**, 163–6

41. Smith, E.M., Anthony, F.W., Gadds, C. and Masson, G.M. (1989). Trial of support treatment with human chorionic gonadotropin in the luteal phase after Buserelin and human menopausal gonadotropin in women taking part in an *in vitro* fertilization program. *Br. Med. J.*, **298**, 1483–6

13

Multiple follicular maturation for assisted reproduction

S.M. Walker

INTRODUCTION

In 1978 Edwards and Steptoe[1] reported the first live birth as a result of *in vitro* fertilization (IVF) using the natural cycle and so brought the chance of pregnancy closer for many couples. Prior to this the technique had been attempted by employing ovarian stimulation regimens similar to those used successfully in animal husbandry. Early failures were attributed to disturbance of the luteal phase by the abnormal production of ovarian steroids as a consequence of hyperstimulation. Little success attended the efforts of other workers to repeat that of Edwards and Steptoe until the Australian group, under the direction of Trounson and colleagues[2], reverted to the use of ovarian stimulation. Such regimens provided an increased number of oocytes and also, consequent upon the administration of human chorionic gonadotrophin (hCG), diminished the then current problem of determining the initial surge of luteinizing hormone (LH) which heralds ovulation. This enabled more accurate timing of follicle aspiration for oocyte retrieval with an increased pregnancy rate attributable to the transfer of more than one embryo.

NATURAL OR STIMULATED CYCLES

Since the early 1980s, comparative studies of natural and stimulated cycles have consistently shown a higher pregnancy rate in association with the stimulated cycle and this has led to the evaluation of varying combination and single drug regimens[3-6]. The effect of using anti-oestrogens and gonadotrophins singly or in combination is to increase the interval during which follicle stimulating hormone (FSH) remains above the threshold level during an individual ovarian cycle, allowing for further recruitment after the second wave (cohort) of follicles. As suggested by Baird[7] the width of the FSH 'gate' determines the number of follicles which will be included in the selection process and subsequently enter the growth and maturation phase.

To compensate, therefore, for the low pregnancy rate obtained in women receiving natural cycle IVF, endeavours have been directed over the last decade towards further understanding human reproduction and, in parallel, devising drug regimens to enable more mature oocytes to be retrieved and consequently more than a single embryo to be available for *in utero* transfer. The restriction of transfer of up to three embryos (exceptionally four) imposed by the Human Fertilisation and Embryology Authority[8] minimizes the chance of high-order multiple pregnancy with the attendant risks of prematurity and perinatal death.

MONITORING OF CYCLES

Close monitoring of ovarian response is generally undertaken by ultrasonography, including, in some instances, estimation of LH and peripheral oestrogen concentrations. Although the value of hormonal monitoring for IVF remains contentious, many favour its inclusion as an additional guide to clinical management[5,9], the ultimate aim of which is to obtain at least one mature oocyte capable of being fertilized and subsequently developing into normal offspring. It had been suggested by Trounson and Mohr[10], in the early 1980s, that an association exists between peak peripheral plasma oestrogen levels and the number of follicles with a diameter $>15\,mm$ (determined by ultrasound), from each of which an oocyte is expected to be obtained. However, early experience with the newer preparations, and particularly recombinant FSH, has revealed that follicular growth can occur with a lower oestradiol production than hitherto experienced and the currently accepted ratio of approximately $0.8\,nmol/l$ ($300-500\,pg/ml$) per potential oocyte may need revision.

The realization that IVF pregnancy rates per treatment cycle may be augmented by the transfer of frozen–thawed embryos derived from the original stimulation[11] has the potential to further encourage the promotion of regimens specifically aimed at achieving multiple follicle maturation.

SUPEROVULATION REGIMENS

From experiences with IVF, superovulation has become a fundamental component of various assisted reproduction programmes in recent years; this is influenced principally by the ultimate technique to be employed, e.g. gamete intrafallopian transfer, intrauterine insemination. Thus, superovulation associated with intrauterine insemination is advocated by many units as the appropriate treatment for couples exhibiting unexplained infertility. Variable pregnancy rates have been reported[12,13], a significant feature being the differing drug regimens employed. In a study of couples with unexplained infertility, which was undertaken in the University Hospital of Wales using a combination of pituitary desensitization and gonadotrophin stimulation with intrauterine insemination, the pregnancy rate achieved was 40.6% per treatment cycle. The superiority of this protocol is due, we believe, to the precision achieved. We have also applied this regimen to our donor insemination programme with similar success (26% per treatment cycle and a cumulative pregnancy rate over 3 months of 55%). To minimize the risk of multiple pregnancy it is imperative to adhere to strict criteria. Within this treatment modality particularly, the combination of ultrasound and hormone monitoring is invaluable. Regardless of the nature of ovarian stimulation, insemination should be withheld and coitus discouraged if more than three follicles >18 mm diameter (from ultrasound examination) are present, associated with oestrogen levels which are suggestive of an increased risk that multiple pregnancy may supervene.

Experience with many combinations of drugs has highlighted the relative unpredictability of response after the first application, despite taking into account such factors as menstrual cycle length, age and pre-existing ovarian dysfunction. The majority of women, however, will respond within a predetermined range to a relatively small number of established combination regimens.

Clomiphene and hMG

Until relatively recently the most commonly employed regimen was clomiphene citrate (CC) in combination with human menopausal gonadotrophin (hMG), with the addition of hCG when the lead follicle had achieved a mean ultrasound diameter of 16–18 mm.

Early studies such as those undertaken by Lopata[5] demonstrated that the pregnancy rate achieved with CC+hMG exceeded that in women treated with CC alone. Analysis of data indicated that the use of hMG in combination with CC resulted in a larger average number of oocytes retrieved, the addition of hMG sustaining the growth of those follicles recruited by the effect of CC. When considering fertilization and cleavage rates, however, the proportion retrieved was significantly greater with clomiphene citrate alone. Vargyas and Marrs[4], in a similar study, were able to show that the increase was due to the presence of immature oocytes, and noted that the fertilization rate correlated inversely with the dose of CC+hMG. These findings suggest that synchronous nuclear and cytoplasmic development of the oocyte is more readily achieved with lower levels of stimulation. Achieving a pregnancy is dependent upon the quality of the oocyte in terms of synchronous maturation of its component parts, in addition to other factors such as semen quality, which influence fertilization, and endometrial receptivity.

Newer modifications

It has become increasingly evident over the last decade that attempts to further improve pregnancy rates by manipulation of super-stimulation regimens were foundering not only because of variable oocyte maturity and quality at the time of retrieval, but also because of an associated effect on endometrial receptivity, either directly or in relation to a derangement of oestrogen:progesterone ratio.

Recent modification and development of new generations of drugs has been directed specifically towards improving oocyte and therefore embryo quality, and minimizing the disruptive effect of ovarian steroids on endometrial receptivity. The assessment of oocyte quality beyond that of grading maturation morphologically, however, remains a practical clinical problem.

One may argue that current philosophy should be to improve ovarian stimulation regimens such that the naturally occurring species quota for the human female may be approached more closely. This would itself minimize the risks of excessive ovarian steroid production, whether directed towards the uterine environment or to the woman as a whole. Ovarian folliculo-genesis beyond the small antral stage has been shown to be dependent upon cyclic gonadotrophin stimulation – principally FSH – producing a species-specific quota of recruited follicles destined to progress either to maturation or atresia during a single cycle. In the human female the natural quota is said to be 6–30 antral follicles, only one of which is usually destined to ovulate.

GnRH agonists and *in vitro* fertilization

An important association between LH and both oocyte and embryo quality has been recognized, in addition to that of the long-standing spatial relationship between LH and ovulation. Reports have been published[14,15] describing the results of extensively monitored cycles both in IVF programmes and in relation to ovulation induction which incriminate elevated LH during the late follicular phase as a cause of poor-quality oocytes and embryos, with the ultimate adverse effect on pregnancy rates. Such a pathological elevation of LH is also described in some women with polycystic ovaries[16].

With the introduction of gonadotrophin releasing hormone agonists (GnRH-a), cancellation rates which previously had been as high as 30–40% per cycle have been reduced to as low as 5%, principally due to the virtual elimination of the rise in endogenous LH production. Many Units have recorded an increase in the number of oocytes obtained (with an associated improvement in pregnancy rates per cycle commenced and per embryo transfer (ET)) following implementation of a combination regimen comprising GnRH-a and gonadotrophins, particularly those National Health Service centres which require considerable flexibility with regard to the timing of oocyte retrieval.

GnRH agonists are principally employed in three modes, comprising ultrashort, short and long protocols. Immediately prior to pituitary desensitization, administered GnRH-a induces a so-called 'flare effect', whereby the discharge of endogenous FSH induces an increased recruitment of small follicles. This phenomenon has a potentially adverse effect if it occurs during the preceding luteal phase, as with the short protocol. Under these circumstances it induces the development of a number of small follicles, exhibiting a reduced potential for developmental maturity but with the capacity for contributing to the overall size of the ovarian pool, both physically and hormonally. Cystic structures reported[17] as being present in as many as 20% cycles at the onset of ovarian stimulation may have originated during the previous luteal phase and been rescued from atresia by the discharge of gonadotrophin or during the early follicular phase of the preceding cycle. The presence of persistent follicular structures appears not to adversely influence outcome, with regard to oocyte retrieval, although because they are oestrogen-rich, this can contribute to confusion when interpreting the results of monitoring. Simple ultrasound-guided aspiration prior to commencing ovarian stimulation may therefore prove useful. The presence of persistent cystic corpora lutea may be effectively eliminated using the technique of menstrual delay.

To be effective, GnRH-a needs to be given subcutaneously in both the short and ultra-short protocols, the latter covering the early recruitment phase and therefore supplementing the effects of administered gonadotrophins. The long protocol, which relies upon achieving ovarian suppression before commencing stimulation regimens, is of particular value for those women with a tendency to hyperstimulate (i.e. with polycystic ovaries (PCOs)), and the older woman for whom treatment may be prolonged.

Reported pregnancy rates are variable when comparing gonadotrophin therapy with or without GnRH-a. Tur-Kaspa and colleagues[18] described a delay of 1.5 days in estimated implantation in 40% of women receiving GnRH-a + hMG compared with hMG alone based on elevated βhCG. They also noted an associated increase in overall pregnancy rate which, it was suggested, was due to the combination of GnRH-a and hMG widening the implantation window, allowing greater opportunity for successful implantation. Further evidence supporting this concept was expressed by Younis and colleagues[19] in describing their experience with hormone therapy prior to oocyte donation.

Other workers doubt the value of GnRH-a for routine inclusion in treatment protocols, preferring to reserve it for those women with polycystic ovaries. Polson and colleagues[20] have recently suggested that it should only be used for women with endogenous LH surge or a poor response to CC + hMG in a previous treatment cycle.

Rutherford and colleagues[21] and Macnamee and co-workers[22] have independently shown an increase in the 'take home baby rate' for those women receiving gonadotrophin therapy and GnRH-a. In addition to improving embryo quality, it is postulated that the oestrogen-sparing effects of GnRH-a significantly minimize disruption of the endometrium and favour implantation.

Overstimulation

In addition to asynchrony in the development of oocytes and endometrium contributing to early pregnancy failure, a major, life-threatening complication of multiple follicular maturation is ovarian hyperstimulation syndrome (OHSS). Although ultrasound monitoring of ovarian response to stimulation is able to determine the number and size of follicles it cannot exclude the development of OHSS during treatment or early pregnancy. Peripheral oestradiol concentrations provide a useful but limited additional guide[23], as high pre-retrieval levels are not pathognomonic for OHSS. In general terms, however, a high oestradiol concentration (> 2000 pg/ml) in conjunction with numerous medium size follicles is indicative of an increased risk of developing moderate to severe OHSS. If a high risk is recognized, there are a number of strategies which may be appropriately adopted to minimize the chance of developing the overt syndrome. A commonly employed method is to puncture all accessible follicles at the time of oocyte retrieval and then freeze any embryos for transfer during a subsequent cycle. If intranasal GnRH-a is being used, it is beneficial to continue the agonist treatment and withhold further gonadotrophin stimulation until the follicles have subsided. Following suppression of the adverse effects of superstimulation, gonadotrophins may be successfully reintroduced at a lower dose. Importantly, hCG administration should be withheld, as OHSS is principally observed in the presence of hCG, whether exogenous or produced as a result of ectopic or intrauterine pregnancy. Luteal support may be administered in such patients, if required, in the form of progesterone.

A further alternative strategy may be to proactively reduce the level of gonadotrophin stimulation if there are indications of excessive ovarian response, thus narrowing the FSH window and allowing the smaller follicles to become atretic, meanwhile sustaining the growth of the dominant follicles. Schoot and colleagues[24], from the Netherlands, recently reported a reduction in oestradiol and FSH levels by 26% 3 days after reducing the dose of hMG in a group of women with polycystic ovaries undergoing induction of ovulation. We have applied this technique to women undergoing ovarian stimulation for IVF and found a similar response, the magnitude of which appears to be related to follicle size immediately prior to dose reduction. It is noteworthy that this strategy is accompanied by symptomatic improvement, although the data are as yet incomplete.

Women with polycystic ovarian syndrome (PCOS) have been shown to have an increased risk of developing OHSS, and the balance between obtaining adequate follicular growth to enable retrieval of appropriately mature oocytes and the prevention of OHSS is difficult to achieve. Attempts to minimize the risk of severe OHSS by manipulating exogenous gonadotrophin stimulation to levels near the ovarian threshold from the outset, with the express purpose of maintaining growth of the larger first-wave follicles (thus allowing atresia of medium and small follicles which have failed to achieve critical-level FSH responsiveness) is a useful approach which requires further investigation. Similarly, the use of GnRH-a instead of hCG to induce oocyte maturation should be explored further[25].

The principles of prevention of moderate to severe OHSS therefore centre upon the recognition of high-risk groups, such as those with PCOS, and the overall reduction of the functional ovarian unit. Despite generalized enlargement and increased vascularity of the ovary, little has been published regarding the risks of torsion and intracystic haemorrhage in the absence of OHSS, during ovarian stimulation for assisted reproduction.

SUMMARY

The probability of achieving a pregnancy in assisted reproduction is related to the number of mature follicles obtained and consequently to the quality of gametes and embryos transferred. In addition, importance is attached to uterine receptivity at the level of the endometrium. Non-invasive methods are being developed in an attempt to determine good, universally applicable prognostic indicators. These include the measure-ment of flow velocity waveforms for the vessels which supply specific structures, such as the ovarian follicle and uterus, and functional evaluation of the embryo–corona cell complex. We suggest that the ultimate aim should be to develop optimal ovarian stimulation regimens which will achieve a pregnancy as a result of the transfer of one, or two, embryos, whilst minimizing the chance of multiple pregnancy, and eliminating derangement in maturation of oocytes and endometrium which predispose to pregnancy loss including ectopic pregnancy. The need for 'spare' oocytes should be addressed as a separate issue.

REFERENCES

1. Steptoe, P.C. and Edwards, R.G. (1978). Birth after the re-implantation of a human embryo (letter). *Lancet*, **2**, 366
2. Trounson, A.O., Leeton, J. F., Wood, C., Webb, J. and Wood, J. (1981). Pregnancies in humans by fertilisation *in vitro* and embryo transfer in the controlled ovulatory cycle. *Science*, **212**, 681–2
3. Jones, H.W., Jones, G.S. and Andrew, M.C. (1982). The programme for *in vitro* fertilisation in Norfolk. *Fertil. Steril.*, **38**, 14–21
4. Vargyas, J.M. and Marrs, R.R. (1984). The effect of different methods of ovarian hyperstimulation and methods of monitoring ovarian response for IVF and ET. *J. IVF and ET*, **1**, 56
5. Lopata, A. (1983). Concepts in human *in vitro* fertilisation and embryo transfer. *Fertil. Steril.*, **40**, 289–301
6. Steptoe, P.C., Edwards, R.G. and Walters, D.E. (1986). Observation on 767 clinical pregnancies and 500 births after human *in vitro* fertilisation. *Hum. Reprod.*, **1**, 89–94
7. Baird, D.T. (1987). A model for follicular selection and ovulation: lessons from superovulation. *J. Steroid Biochem.*, **27**, 15–23
8. Human Fertilisation and Embryology Act (1990)
9. Fischel, S.B., Edwards, P.G. and Purdy, J.M. (1984). Analysis of 25 infertile patients treated consecutively by *in vitro* fertilisation at Bourn Hall. *Fertil. Steril.*, **42**, 191–7
10. Trounson, A.O. and Wood, C. (1981). Extracorporeal fertilisation and embryo transfer. *Clin. Obstet. Gynecol.*, **8**, 681–713
11. Trounson, A.O. and Mohr, L. (1983). Human pregnancy following cryopreservation, thawing and transfer of an 8-cell embryo. *Nature (London)*, **305**, 707–9

12. Dodson, W.C. and Haney, A.F. (1991). Controlled ovarian hyperstimulation and intrauterine insemination for treatment of infertility. *Fertil. Steril.*, **55**, 457–67

13. Iffland, C.A., Reid, Amso, N., Bernard, A.G., Buckland, G. and Shaw, R.W. (1991). A within-patient comparison between superovulation with intra-uterine artificial insemination using husband's washed spermatozoa and gamete intrafallopian transfer in unexplained infertility. *Eur. J. Obstet. Gynaecol. Reprod. Biol.*, **39**, 181–6

14. Macnamee, M.C., Edwards, R.G. and Howles, C.M. (1988). The influence of stimulation regimes and luteal phase support on the outcome of IVF. *Hum. Reprod.*, **3**, (Suppl. 2) 43–52

15. Regan, L., Owen, E.J. and Jacobs, H.S. (1990). Hypersecretion of luteinising hormone, infertility and miscarriage. *Lancet*, **336**, 1141–4

16. Homburg, R., Armar, N.A., Eshel, A., Adams, J. and Jacobs, H.S. (1988). Influence of serum luteinising hormone concentration on ovulation, conception and early pregnancy loss in polycystic ovary syndrome. *Br. Med. J.*, **297**, 1024–6

17. Ben-Rafael, Z., Bider, D., Menashe, Y., Mayman, R., Zolti, M. and Mashiach, S. (1990). Follicular and luteal cysts after treatment with gonadotrophin releasing hormone analog for *in vitro* fertilisation. *Fertil. Steril.*, **53**, 1091–4

18. Tur-Kaspa, I., Confino, E. and Dudkiewicz, A.B. (1990). Ovarian stimulation protocol for *in vitro* fertilisation with gonadotrophin releasing hormone agonist widens the implantation window. *Fertil. Steril.*, **53**, 859–65

19. Younis, J.S., Mordel, N., Ligovetzkyg, Lewin, A., Schenker, J.G. and Laufer, N. (1991). The effect of prolonged artificial follicular phase on endometrial development in an oocyte donation programme. *J. IVF and ET*, **8**, 84–8

20. Polson, D.W., MacLachlan, V., Krapez, J.A., Wood, C. and Healey, D. (1991). A controlled study of gonadotrophin-releasing hormone agonist (buserelin acetate) for folliculogenesis in routine *in vitro* fertilisation patients. *Fertil. Steril.*, **56**, 509–14

21. Rutherford, A.S., Suback-Sharpe, R.J., Dawson, K.J., Margara, R.A., Franks, S. and Winston, R.M.L. (1988). Improvement of *in vitro* fertilization after treatment with buserelin, an agonist of luteinizing hormone releasing hormone. *Br. Med. J.*, **296**, 1765–8

22. Macnamee, M.C., Howles, M.C., Edwards, R.G., Taylor, P.J. and Elder, K.T. (1989). Short-term luteinising hormone-releasing hormone agonist treatment, prospective trial of a novel ovarian stimulation regimen for *in vitro* fertilization. *Fertil. Steril.*, **52**, 264–9

23. Delvigne, A., Vandromme, J., Barlow, P., Lejeune, B. and Leroy, F. (1991). Are there predictive criteria of complicated ovarian hyperstimulation in IVF? *Hum. Reprod.*, **6**, 959–62

24. Schoot, D.C., deJong, F.H., Pache, T.D., Fayser, B.C.J.M. and Hop, W.C. (1992). Growth patterns of ovarian follicles during induction of ovulation with decreasing doses of human menopausal gonadotrophin following presumed selection in polycystic ovary syndrome. *Fertil. Steril.*, **57**, 1117–20

25. Itskovitz, J., Boldes, R., Levron, J., Erlik, Y., Kahana, L. and Brandes, J.M. (1991). Induction of preovulatory luteinising hormone surge and prevention of ovarian hyperstimulation syndrome by gonadotrophin releasing hormone agonist. *Fertil. Steril.*, **56**, 213–20

14

Risks associated with ovulation induction

N. Amso

INTRODUCTION

The strict definition of ovulation induction would imply the administration of drugs to stimulate follicular development, and usually of further drugs to induce release of the oocytes. Women with anovulatory infertility must undergo induction of ovulation if they wish to become pregnant. When attempting to induce ovulation in these anovulatory women, the objective is to achieve ovulation of a few (1–3) mature oocytes and the challenge is to imitate the normal cycle as closely as possible. This approach is directed at achieving ovulation and pregnancy in many patients, whilst preventing multiple follicular growth, multiple pregnancies and hyperstimulation.

At present, it is now almost universally accepted that multiple follicular development is essential for a successful assisted reproduction programme. This is mainly due to the finding that the incidence of pregnancy increases in proportion to the number of gametes or embryos replaced, and the more embryos available for cryopreservation the better are the chances of subsequent survival and replacement. The conceptual idea of assisted reproduction techniques is to use a super-physiological dosage of gonadotrophins in order to obtain a large number of eggs for fertilization and transfer.

Over the past decade, a number of drugs have been employed to induce ovulation or to achieve multiple follicular development (Table 1)

Table 1 Drugs commonly used for ovulation induction; hMG, human menopausal gonadotrophin; LH, luteinizing hormone; FSH, follicle stimulating hormone

---

Anti-oestrogens (clomiphene citrate, tamoxifen)

Menopausal gonadotrophins (hMG; LH : FSH = 1 : 1)
– alone
– in conjunction with anti-oestrogens, or

'Pure' follicle stimulating hormone
– alone
– in conjunction with hMG

Gonadotrophin releasing hormone (pulsatile GnRH)

Gonadotrophin releasing hormone agonist (GnRH-a)
– with hMG alone, or
– with FSH and hMG

---

depending on the clinical indications and the preference of any particular centre. The drugs commonly used include anti-oestrogens (clomiphene citrate, tamoxifen), gonadotrophin releasing hormone (GnRH), menopausal gonadotrophins, pure follicle stimulating hormone (FSH) and, more recently, gonadotrophin releasing hormone agonists (GnRH-a) with menopausal gonadotrophins with or without FSH. Inevitably, such treatments are associated with a number of complications resulting from the medication itself, or as a consequence of the ovulation induction or assisted reproduction treatment. It is imperative that clinicians should define the risks as those not only associated with the medication itself but also those short- or long-term effects which are either physical, emotional or pregnancy-related. This chapter will review the published literature related to these risks.

DRUG-RELATED RISKS

Minor side-effects resulting from the pharmacological effects of the above drugs do occur, but they rarely interfere with the treatment. These will be discussed in relation to each individual drug.

Clomiphene citrate

Approximately 10% of women complain of hot flushes while on clomiphene citrate treatment and less than 2% complain of other minor side-effects such as nausea, vomiting, breast tenderness, hair loss or visual disturbances[1]. In addition, concern has been raised about the effects of clomiphene on oocytes and embryos. Large surveys of women who conceived whilst on clomiphene reported no significant increase in congenital malformation rates[2]. Recently, however, there has been some concern over an increase in the incidence of neural tube defects in the offspring of women who have undergone ovulation induction[3]. However, others[4] have suggested that the present evidence is insufficient to judge whether the association is real and that it is relatively safe to both patients and offspring. Nevertheless the final conclusions with regard to long-term side-effects must await sufficiently large surveys of post-pubertal offspring[5].

Gonadotrophins and pulsatile GnRH

There are few reports of drug-associated side-effects related to exogenous administration of gonadotrophins. Children born after such treatment have not shown any increased incidence of genetic abnormalities and their reproductive ability appears to be normal.

Administration of pulsatile GnRH may be associated with infection at the cannula site and the subsequent rare development of septicaemia, antibody formation or desensitization. GnRH-specific immunoglobulin G and E antibodies to GnRH in men and women on long-term treatment with pulsatile GnRH have been reported. Desensitization may arise as a result of an inadequate dose or pulse frequency of GnRH, impaired absorption, or with excessive GnRH. Both pituitary and gonadal desensitization have been observed in monkeys and humans with GnRH given as a continuous infusion[6].

Gonadotrophin releasing hormone agonist

The use of GnRH-a may be associated with side-effects due to hypo-oestrogenaemia. The risk of the agonist-inducing immunization against itself appears to be minimal and in humans, no antibody production against buserelin (Hoechst, Germany) has been detected, even after prolonged use. In assisted reproduction programmes, inadvertent pregnancies have been reported after starting GnRH-a at mid-cycle or in the mid-luteal phase[7–9]. So far, there are no reports of major congenital abnormalities in births following early exposure to GnRH-a and it would be tempting to suggest that in these patients with a long history of infertility, the outcome might justify the risk. However, it is crucial to await the results of long-term follow-up studies to determine the full impact of such early exposure on these children. Equally important is the adverse luteolytic effect of continued GnRH-a administration[10] on the progression of early pregnancy.

RISKS ASSOCIATED WITH OVULATION INDUCTION OR GAMETE/EMBRYO REPLACEMENT

The incidence of some of these risks is largely dependent on the ovulation induction protocol being employed, dosage and duration of medication, and diligence in the monitoring of the treatment cycle. These risks include ovarian hyperstimulation syndrome, multiple pregnancy, psychological and emotional effects and potential risk of cancer.

Ovarian hyperstimulation syndrome (OHSS)

This is a serious complication of ovulation induction. Its incidence varies considerably and is much more common following gonadotrophin administration than clomiphene treatment. The management of OHSS depends on the stage at which the diagnosis is made. Cancellation of the treatment cycle by withholding the human chorionic gonadotrophin (hCG) injection generally prevents the development of OHSS and might be preferable under certain circumstances. However, it is not always possible to predict the occurrence of OHSS before the hCG injection and furthermore, oocytes with a potential to produce viable embryos are wasted when the treatment cycle is aborted. When an embryo cryopreservation programme is available, the oocytes may be collected, fertilized and all the developing embryos cryopreserved for transfer at a later date. Good results have been achieved with this approach[11–13]. Other alternative policies have been proposed for the management or prevention of predicted OHSS in women undergoing GnRH-a/gonadotrophin stimulation[14].

Multiple pregnancy

A multiple pregnancy rate of 6–7% has been reported after ovulation induction with clomiphene[15]. Clomiphene in combination with hCG is said to be associated with an increased multiple pregnancy rate and, while the majority are twin pregnancies, triplets, quadruplets and quintuplets have been reported. Following gonadotrophin treatment, 20–38% of pregnancies may be multiple, in spite of meticulous monitoring with hormone assays and ultrasound scans. The incidence is dependent on the total number of developing follicles and the number of mature follicles at the time of hCG administration. In contrast, the use of pulsatile GnRH is associated with a lower rate of multiple pregnancy, averaging approximately 5% (range 0–14%)[6]. Women with hypogonadotrophic hypogonadism undergoing their first treatment cycle with pulsatile GnRH have an increased rate of multiple pregnancy[16] while those with polycystic ovarian disease appear to have a low rate. With assisted reproduction treatment, the incidence of multiple pregnancy depends on the number of gametes/embryos replaced ranging from 1–31%[17].

The risks associated with multiple pregnancy are considerable and extend throughout pregnancy and after delivery. Premature birth, low birth weight, and the need for long periods of special neonatal care with possible long-term effects add further complications to the treatment (Table 2)[18]. Such concerns have led to the introduction of limitations on the number of oocytes/embryos replaced to reduce the incidence of high-order multiple births.

Psychological and emotional effects

The intensive monitoring that is commonly required during ovulation induction may cause undue stress and anxiety. In the immediate aftermath of a failed treatment cycle or following early pregnancy loss, the couple may experience shock, frustration and depression. Long-term effects of

Table 2 Influence of multiple births on perinatal mortality and morbidity of infants resulting from *in vitro* fertilization (IVF) or gamete intrafallopian transfer (GIFT). Data from the Medical Research Council (MRC) IVF register[18]

	Singleton	*Twin*	*Triplets and higher-order*	*Total*
Percentage weighing < 2500 g	12	55	94	32
Percentage in special care baby units	14	41	89	29
Stillbirth rate (per 1000)	5.3	18.8	30.5	12.0
Perinatal death rate (per 1000)	11.7	39.7	79.3	27.2
Infant death rate (per 1000)	10.7	34.1	69.2	23.7

failed treatment(s) may lead to marital discord and breakdown of the relationship. Similarly, the psychological difficulties encountered by mothers after the birth of twins or triplets are enormous. The increased perinatal mortality and morbidity of infants resulting from assisted reproduction and summarized in Table 2 are well known and further increase the couples' anxiety and emotional strain. Moreover, the majority of these mothers report considerable fatigue and stress, social isolation, strain on marital relationship, emotional detachment, and difficulties in their relationship with their children[19]. The provision of increased help, counselling and support is essential to minimize these effects.

Potential risk of cancer

Considerable concern has been voiced lately on the potential risk of cancer from follicular stimulation[20]. Excessive oestrogen secretion has been implicated in ovarian, endometrial, and breast carcinoma, and excessive gonadotrophin secretion has been implicated in ovarian cancer. Several reports of women developing carcinoma of the breast, ovaries or the Fallopian tube following ovulation induction have appeared in the literature. Although these cases do not prove a causal link between ovarian stimulation and genital cancer, gonadotrophins or other associated factors may be an added stimulus, and ovarian hyperactivity may accelerate ovarian neoplasia. Long-term studies are essential in this respect.

SUMMARY

The risks associated with ovulation induction are multiple and varied. The final conclusions with regard to the long-term risks of a number of these medications, on women or their offspring, must await results of large follow-up surveys. Attention to the wider and long term implications should lead clinicians to offer couples routine counselling and to exercise extreme care and caution before and during treatment.

REFERENCES

1. Kistner, R.W. (1968). Induction of ovulation with clomiphene citrate (clomid). *Obstet. Gynecol. Surv.*, **20**, 873–900
2. McKenna, K.M. and Pepperell, R.J. (1988). Anti-oestrogens: their clinical physiology and use in reproductive medicine. *Bailliere's Clin. Obstet. Gynaecol.*, **2**, 545–6
3. Czeizel, A. (1989). Ovulation induction and neural tube defects. *Lancet*, **2**, 167
4. Cuckle, H. and Wald, N. (1989). Ovulation induction and neural tube defects. *Lancet*, **2**, 1281
5. Lunenfeld, B., Blankstein, J., Koter-Emeth, S., Kokia, E. and Geier, A. (1986). Drugs used in ovulation induction. Safety of patients and offspring. *Hum. Reprod.*, **1**, 435–9
6. Shoham, Z., Homburg, R. and Jacobs, H.S. (1990). Induction of ovulation with pulsatile GnRH. *Bailliere's Clin. Obstet. Gynaecol.*, **4**, 589–608
7. Martinez, F., Barri, P.N. and Coroleu, V. (1988). Accidental GnRH agonist administration during early pregnancy. *Hum. Reprod.*, **3**, 669
8. Ron El, R., Golan, A., Herman, A., Raziel, A., Soffer, Y. and Caspi, E. (1990). Midluteal gonadotropin-releasing hormone analog administration in early pregnancy. *Fertil. Steril.*, **53**, 572–4
9. Jackson, A.E., Curtis, P., Amso, N. and Shaw, R.W. (1992). Exposure to LHRH agonists in early pregnancy following the commencement of mid-luteal buserelin for IVF stimulation. *Hum. Reprod.*, **7**, 1222–4

10. Casper, R.F. and Yen, S.S.C. (1979). Induction of luteolysis in the human with a long-acting analog of luteinizing hormone releasing factor. *Science*, **205**, 408–410

11. Amso, N.N., Ahuja, K.K., Morris, N. and Shaw, R.W. (1990). The management of predicted ovarian hyperstimulation involving gonadotropin-releasing hormone analog with elective cryopreservation of all pre-embryos. *Fertil. Steril.*, **53**, 1087–90

12. Amso, N. and Shaw, R.W. (1991). Polycystic ovaries and problems in assisted reproduction programmes. In Shaw, R.W. (ed.) *Polycystic ovaries: a disorder or a symptom?*, pp. 203–16. (Carnforth: Parthenon Publishing)

13. Rizk, B., Davies, M., Steer, S., Bell, S.C., Dillon, D. and Edwards, R.G. (1992). The immunohistochemical expression of endometrial proteins and pregnancy outcome in frozen embryo replacement cycles. *Hum. Reprod.*, **7**, 413–17

14. Forman, R.G., Frydman, R., Egan, D., Ross, C. and Barlow, D.H. (1990). Severe ovarian hyperstimulation syndrome using agonists of gonadotropin-releasing hormone for *in vitro* fertilization: a European series and a proposal for prevention. *Fertil. Steril.*, **53**, 502–9

15. World Health Organization (1973). Agents stimulating gonadal function in the human. *Technical report series no. 153*, p.15. (Geneva: WHO)

16. Homburg, R., Eshel, A., Armar, N.A., Tucker, M., Mason, P.W., Adams, J., Kilborn, J., Sutherland, I.A. and Jacobs, H.S. (1989). One hundred pregnancies after treatment with pulsatile luteinising hormone releasing hormone to induce ovulation. *Br. Med. J.*, **298**, 809–12

17. Voluntary Licensing Authority. (1990). The Fifth Report of the Voluntary Licensing Authority for Human *In Vitro* Fertilisation and Embryology. (London)

18. Medical Research Council working party on children conceived by *in vitro* fertilisation (1990). Births in Great Britain resulting from assisted conception, 1978–1987. *Br. Med. J.*, **300**, 1229–33

19. Garel, M. and Blondel, B. (1992). Assessment at 1 year of the psychological consequences of having triplets. *Hum. Reprod.*, **7**, 729–32

20. Fishel, S. and Jackson, P. (1989). Follicular stimulation for high tech pregnancies: are we playing it safe? *Br. Med. J.*, **299**, 309–11

Index

adrenal hyperandrogenaemia 59–60
amenorrhoea
    exercise-related 55–56
    weight loss-related 54–55
assisted reproduction
    multiple follicular maturation for
        151–161

biopsy, endometrial
    in assessment of corpus luteal
        function 138
body temperature, basal
    in assessment of corpus luteal
        function 137

cancer
    potential risk of, in ovulation
        induction 168–169
childhood
    follicular maturation in 35–48
clomiphene citrate
    and luteal phase insufficiency in
        ovulation induction 142–143
    in treatment of luteal phase defects
        141–142
    risks associated with, in ovulation
        induction 165
corpus luteal insufficiency 137–149
    corpus luteal function, assessment
        of 137–138
        basal body temperature 137

endometrial biopsy 138
    serum progesterone 138
    evaluating, problems in 139
    gonadotrophins and GnRH
        agonists 144–145
    luteal phase, in IVF 145–146
    luteal phase, in ovulation induction
        142–144
        clomiphene citrate 142–143
        human menopausal
            gonadotrophins 143
luteal phase defects, treatment of
    140–142
    clomiphene citrate 141–142
    progesterone 140

dominant follicle 14–18

embryo replacement
    risks associated with 166–169
        multiple pregnancy 167
        ovarian hyperstimulation
            syndrome 166–167
        potential cancer risk 168–169
        psychological and emotional
            effects 167–168
embryology, ovarian 35, 36
empty follicle syndrome 24
endocrine
    disorders, ovulatory dysfunction in
        49–66

regulation of follicular oestrogen
 synthesis 30
role in follicular maturation 39–41
endometrial biopsy
 in assessment of corpus luteal
  function 138
exercise-related amenorrhoea 55–56

follicle
 development, concepts of 5–6
 growth of, and gonadotrophins
  1–11
 maturation
  in childhood and puberty 35–48
  multiple, for assisted
   reproduction 151–161
 morphology, preovulatory 13–26
follicle stimulating hormone (FSH)
 heterogeneity 2–5
 natural, comparison with human
  recombinant FSH 1–11
 purified
  advantages and disadvantages of
   68–69
  induction of ovulation 69–71
 recombinant human, clinical
  aspects of 75–86
FSH, recombinant human
 characteristics 1–2
 clinical aspects 75–86
 comparison with natural FSH 1–11

gamete replacement
 risks associated with 166–169
  multiple pregnancy 167
  ovarian hyperstimulation
   syndrome 166–167
  potential cancer risk 168–169
  psychological and emotional
   effects 167–168
gonadotrophin releasing hormone
 (GnRH)

analogues, oocyte postmaturity and
 131–133
pulsatile, in ovulatory disorders
 87–95
 clinical results 91–93
 regimens of administration
  88–91
 risks associated with 165
gonadotrophin releasing hormone
 agonists (GnRH-a)
 and *in vitro* fertilization 155–157
 in corpus luteal insufficiency
  144–145
 risks associated with, in ovulation
  induction 165–166
gonadotrophins
 and follicular growth 1–11
 in corpus luteal insufficiency
  144–145
 preparations, purified, for ovulation
  induction 67–74
 recombinant human follicle
  stimulating hormone (recFSH)
  75–86
 risks associated with, in ovulation
  induction 165
growth hormone
 and ovarian stimulation 97–110
  co-treatment with
   gonadotrophins 98–100
  in IVF–ET 100–105

human chorionic gonadotrophin
 (hCG)
 and recFSH in folliculogenesis 8–9
human menopausal gonadotrophin
 (hMG)
 in ovulation induction
  and luteal phase insufficiency
   143–144
  comparison with FSH 70–71
hyperandrogenaemia, adrenal 59–60

hyperinsulinaemia 59
hyperprolactinaemia 52–53
hyperstimulation, ovarian
  in multiple follicular maturation
    157–158, 166–167
hypogonadotrophic hypogonadism
  pulsatile GnRH in 91

in vitro fertilization (IVF)
  GnRH-a and 155–157
  luteal insufficiency and 145–146
  oocyte morphology and 13
  premature luteinization and
    129–133
in vitro fertilization–embryo transfer
  (IVF–ET)
  growth hormone in 100–105
induction of ovulation
  gonadotrophin preparations for
    67–74
  risks associated with 163–170
    drug-related 164–166
insulin-like growth factors (IGFs)
  97–110

luteal phase defects, treatment of
  140–142
  clomiphene citrate 141–142
  progesterone 140
luteinization, premature 125–136
  animal studies 126
  definition 125
  evidence from IVF 129–133
    incidence and complications
      130–131
    oocyte postmaturity and GnRH
      analogues 131–133
  experimental 133–134
  human studies 126
    ovulation induction in PCOS
      128–129
    stimulated cycles 128

unstimulated cycles 126–127
luteinized unruptured follicle (LUF)
  syndrome 21–24
luteinizing hormone (LH)
  and folliculogenesis 1–11

maturation, follicular
  in childhood and puberty 35–48
    clinical disorders of 44–46
    endocrinology of 39–41
    physical signs of 36–38
    radiological evidence of 41–43
  multiple, for assisted reproduction
    151–161
    monitoring of cycles 152–153
    natural or stimulated cycles
      151–152
    superovulation regimens
      153–158
morphology, follicular
  abnormal 21–24
  preovulatory 13–26

oestrogen synthesis, follicular
  endocrine regulation of 30
  paracrine regulation of 31, 32
    granulosa-derived 31
    in vivo evidence for 32
    theca-derived 31
ovarian failure
  primary 49–52
    general features 49–50
    pathophysiology 50–52
  secondary 52–56
    exercise-related amenorrhoea
      55, 56
    general features 52
    hyperprolactinaemia 52, 53
    weight loss-related amenorrhoea
      54, 55
ovarian function
  local control of 27–34

ovarian hyperstimulation syndrome (OHSS)
  in multiple follicular maturation 157–158, 166–167
ovarian stimulation
  growth hormone and 97–110
ovarian surgery 111–124
ovary, embryology of 35, 36
overstimulation
  in superovulation for assisted reproduction 157–158
ovulation induction
  gonadotrophin preparations for 67–74
  luteal phase insufficiency in 142–144
  risks associated with 163–170
    drug-related 164–166
    multiple pregnancy 167
    ovarian hyperstimulation syndrome 166–167
    potential cancer risk 168–169
    psychological and emotional effects 167–168
ovulatory dysfunction
  in endocrine disorders 49–66
  pulsatile GnRH in 87–95

polycystic ovarian disease 56–60
  adrenal hyperandrogenaemia 59–60
  defined 111–112
  general features 56–58
  GnRH in 91–93
  hyperinsulinaemia 59
  ovulation induction in 128–129
  purified FSH in 69
  surgical treatment for 112–121
precocious puberty 45, 46
pregnancy, multiple
  risk of, in ovulation induction 167
premature luteinization 125–136

animal studies 126
definition 125
evidence from IVF 129–133
  incidence and complications 130–131
  oocyte postmaturity and GnRH analogues 131–133
experimental 133–134
human studies 126
  ovulation induction in PCOS 128–129
  stimulated cycles 128
  unstimulated cycles 126–127
premature thelarche 44, 45
preovulatory follicle
  morphology of 13–26
primary ovarian failure 49–52
  general features 49–50
  pathophysiology 50–52
progesterone
  in treatment of luteal phase defects 140
  serum, in assessment of corpus luteal function 138
puberty
  early and late 45–46
    McCune-Albright syndrome 45
    precocious 45
    Turner's syndrome 46
  follicular maturation in 35–48
purified FSH
  advantages and disadvantages 68–69
  induction of ovulation 69–71
    comparison with hMG 70–71
    efficacy, in polycystic ovary syndrome 69
    low-dose 69–70
recombinant human FSH (recFSH)
  characteristics 1–2
  clinical aspects 75–86

pharmacodynamics 78–83
pharmacokinetics 77–78
safety 76
in folliculogenesis, comparison with
natural FSH 1–11
reproduction, assisted
multiple follicular maturation for
151–161

secondary ovarian failure 52–56
exercise-related amenorrhoea
55–56
general features 52
hyperprolactinaemia 52–53
weight loss-related amenorrhoea
54–55
superovulation 14
regimens for, in assisted
reproduction 153–158

clomiphene and hMG 154
GnRH-a and IVF 155–157
overstimulation 157–158
surgery, ovarian 111–124
current practice 113–114
effectiveness of 114–116
future of 121
history of 112–113
maximizing results 117–118
mode of action 119–121
patient selection for 116–117
safety of 114–116

thelarche, premature 44–45
two-cell, two-gonadotrophin
hypothesis 27–29

weight loss-related amenorrhoea
54–55